THE ALGEBRA OF THOUGHT & REALITY

THE MATHEMATICAL BASIS FOR PLATO'S THEORY OF IDEAS, AND REALITY EXTENDED TO INCLUDE A PRIORI OBSERVERS AND SPACE-TIME

SECOND EDITION

OF

THE ALGEBRA OF THOUGHT & REALITY

A NEW OPERATOR FORMULATION FOR CLASSICAL & QUANTUM LOGIC OBVIATING LOGIC PARADOXES & GÖDEL'S UNDECIDABILITY THEOREM; AND GIVING A MATHEMATICAL BASIS FOR PLATO'S THEORY OF IDEAS, AND REALITY – THE STANDARD MODEL OF PARTICLES

STEPHEN BLAHA, PH.D.

Some Other Books by Stephen Blaha

The Metatheory of Physics Theories, and the Theory of Everything as a Quantum Computer Language (ISBN: 097469584X, Pingree-Hill Publishing, Auburn, NH, 2005)

A Complete Derivation of the Form of the Standard Model With a New Method to Generate Particle Masses SECOND EDITION (ISBN: 9780981904900, Pingree-Hill Publishing, Auburn, NH, 2008)

The Origin of the Standard Model: The Genesis of Four Quark and Lepton Species, Parity Violation, the ElectroWeak Sector, Color SU(3), Three Visible Generations of Fermions, and One Generation of Dark Matter with Dark Energy (ISBN: 0974695882, Pingree-Hill Publishing, Auburn, NH, 2007)

Physics Beyond the Light Barrier: The Source of Parity Violation, Tachyons, and A Derivation of Standard Model Features (ISBN: 0974695874, Pingree-Hill Publishing, Auburn, NH, 2007)

A Derivation of ElectroWeak Theory based on an Extension of Special Relativity; Black Hole Tachyons; & Tachyons of Any Spin (ISBN: 0974695866, Pingree-Hill Publishing, Auburn, NH, 2006)

Quantum Theory of the Third Kind: A New Type of Divergence-free Quantum Field Theory Supporting a Unified Standard Model of Elementary Particles and Quantum Gravity based on a New Method in the Calculus of Variations (ISBN: 0974695831, Pingree-Hill Publishing, Auburn, NH, 2005)

Quantum Big Bang Cosmology: Complex Space-time General Relativity, Quantum Coordinates™Dodecahedral Universe, Inflation, and New Spin 0, ½, 1 & 2 Tachyons & Imagyons™ (ISBN: 0974695815, Pingree-Hill Publishing, Auburn, NH, 2004)

A Unified Quantitative Theory Of Civilizations and Societies: 9600 BC - 2100 AD (ISBN: 0974685858, Pingree-Hill Publishing, Auburn, NH, 2006)

Available on bn.com, Amazon.com, and other web sites as well as at better bookstores (through Ingram Distributors).

Cover Credits
Front cover by Stephen Blaha © 2008, 2009. Part of Raphael's Plato appears in minature.

THE ALGEBRA OF THOUGHT & REALITY

SECOND EDITION

THE MATHEMATICAL BASIS FOR PLATO'S THEORY OF IDEAS, AND REALITY EXTENDED TO INCLUDE A PRIORI OBSERVERS AND SPACE-TIME

STEPHEN BLAHA, PH.D.

Pingree–Hill Publishing

To my wife, Margaret
With Much Love

Er said that the way in which the souls chose their lives was a sight worth seeing, ... The soul of Odysseus made its choice last of all, ... it sought for a long time, looking for the life of a private individual who did his own work.
Plato - The Republic X.620c

PREFACE to the SECOND EDITION

This Second Edition extends the First Edition of *The Algebra Of Thought & Reality: A New Operator Formulation For Classical & Quantum Logic Obviating Logic Paradoxes & Gödel's Undecidability Theorem; and Giving a Mathematical Basis For Plato's Theory Of Ideas, And Reality – The Standard Model Of Particles* in several ways. There are three important new sections. One section discusses Observers both in the formulation of Operator Logic and in the Quantum Reality in which we live. The second new section discusses space-time. It shows the need for Time since, for example, proofs are stated in (time) steps as are experiments and phenomena in Reality. Since we see events at various spatial locations the concept of space must appear in Reality. Consistency with the spinor formulation of Operator Logic leads to four-dimensional space-time. The third section deals with the Concept of Being as substance and form from philosophic and modern particle physics points of view. Lastly, some additional comments appear in the text.

The additional topics presented in this edition serve to solidify the connection of Operator Logic (Ideas-Thought) with Blaha's derivation of the Standard Model (Reality as we currently know it). Thus the chain from Operator Logic to the Standard Model is more solid and based on known entities while other attempts at comprehensive theories of Reality are usually based on unobserved and/or less justifiable constructs, and thus are less compelling.

PREFACE to the FIRST EDITION

Many people have long thought Logic is a product of "pure thought" that is largely independent of our experience of the outside world. Therefore its features have been viewed as universal in the sense that any thinking creature in any part of the universe would arrive at the same concepts of Logic that earth-bound logicians have found.

However, it is clear from the history of Logic that it has evolved considerably in the 19th and 20th centuries from Aristotelian Logic which dominated thought for over two millenia. The work of Frege, Cantor, Boole, Whitehead, Russell, Hilbert, Weyl, Gödel, Turing, Church, and many others of equal note has extended our understanding of Logic enormously. Yet there remains major issues that require clarification, and as this work will show there are major new features, and a new formulation, of Logic that have not been hitherto discovered.

This book describes this new formulation of Logic. It appears to resolve all of the paradoxes that have beset Logic since the 19th century as well as the Liar paradox that dates from early Greek times. It also reduces the importance of Gödel's Undecidability Theorem by showing how to *generally*, and consistently, exclude undecidable propositions from a mathematical-deductive system or its corresponding calculus. The reduced system or calculus then is fully "decidable" – all propositions in the system are either provably true or false. We thus view paradoxes and other undecidable statements as the result of an inadequate (c-number) symbolic/linguistic formulation of Logic that is remedied by an operator (q-number) formulation of Logic.

Remarkably the basis of this new formulation is in the general Measurement Theory of quantum phenomena. The new formulation of Logic uses Quantum Measurement Theory to create an operator (q-number) based Logic using Gödel numbers throughout for signs (symbols). **Thus our deepest knowledge of physical reality is the source of a new, paradox-free, operator formulation of Logic, which we will call *Operator Logic*.** (Operator Logic is not related to Boolean algebra.) Previous formulations of Logic were based on human languages or "intuition" or "everyday physical experience" or "ideal concepts of the human mind" or some combination thereof. Our formulation is rooted in the deepest known bedrock of the physical universe. Thus we achieve a unique foundation for Logic of the greatest depth and magnitude.

It is somewhat surprising to the author that Operator Logic was not developed by either H. Weyl or J. von Neumann, both of whom had a deep knowledge of both Logic and Quantum Theory. It is clear from the standard descriptions of these areas that the relation between Logic and Quantum Measurement Theory was not known to them, or to later specialists in these areas.

Logic can be viewed as rooted in Physics in the sense that our knowledge of Logic is based on experience, and in particular, the general laws of Quantum Measurement Theory.

After developing Operator Logic we show how Plato's theory of Ideas and Reality, and their mathematical relation, is mirrored by the development of the Standard Model of Elementary Particles from the mathematical framework of Operator Logic. This book therefore can be viewed as the precursor of the derivation of the Standard Model given in Blaha (2008). Together these books provide an implementation of Plato's qualitative theoretic framework, which in this author's view is quite remarkable.

This book assumes some knowledge of Logic, matrices, and group theory. Knowledge of Quantum Theory is not required but some knowledge of elementary particles and their interactions is helpful. Quantum concepts are developed as required.

CONTENTS

TABLES & FIGURES

0. Logic

> *What is Truth? ...*
> *No pleasure is comparable to the standing upon the*
> *vantage ground of truth ... and to see the errors,*
> *and wanderings, and mists, and tempests, in the vale below.*
> *"Of Truth" - Francis Bacon*

Logic is the Science of the True and the False. Like a many-faceted diamond it reflects light on itself, on Mathematics, the Sciences, and Philosophy. It has the peculiar nature of being used extensively in Mathematics to derive and prove results, and yet it is often thought of as being the foundation of Mathematics. A number of logicians have devoted their lives to proving that Mathematics can be derived from strictly logical constructs. For example, Frege attempted to derive arithmetic from logical constructs. Subsequently, logicians, such as Russell, showed that Frege's attempt failed. So more ambitious attempts to derive features of Mathematics from Logic, as currently conceived, have little hope of success.

The advent of Quantum Computers, at least in theory, makes the understanding of Logic even more critical since a Quantum Computer can follow many logical paths in parallel in performing a quantum computation.

To develop a sound basis for Logic it is important to understand the source of logical constructs. Where did the constructs "and", "or", "not", and so on originate? In part this is a historical question. Clearly, many arose in the development of human languages over the millenia. But that answer really is not sufficient because it does not explain *why* or *how* these constructs arose.

There are two answers to these questions, and correspondingly, two major schools of thought. One answer is that they arose through "pure thought" – an inherent intuition that mankind possesses. The proponents of this answer are usually called *platonists*, *intuitionists*, or *contensivists*. Their general belief is these constructs inherently exist in some way in the mind.

The other possible answer is that logical constructs are found empirically – perhaps not as the result of experiments – but through practical experience. Thus "and" might have arisen when a "cave man" had to tell two of his children to go for a walk: "Sally *and* Jane go for a walk." The proponents of this view could be called *empiricists*. There is a significant group of logicians and mathematicians that feel mathematics is an empirical science. An interesting piece of support for this view is the "discovery" of quantifiers such as "all" by Frege in the nineteenth century. Here was a construct of an extremely simple nature that had to be introduced into Logic despite 2,500 years of prior work in Logic.

In this book we will generally take the position of empiricists and contend that there are major aspects of Logic, including its fundamental formulation, that have not, until now, been recognized. We will describe a new general formulation of Logic that embodies the somewhat vague conceptualizations of Logic, and its semantic aspects, and that resolves longstanding logical paradoxes as well as issues associated with Gödel's Undecidability Theorem.

We will show that quantum physical experiments can be viewed as logic statements. "Experiments are Statements." Then we will demonstrate that Quantum Measurement Theory has a close (and unstated) relationship with the new operator formulation of Logic presented herein.

This new formulation applies to classical logic and quantum probabilistic logic.

In Blaha (2005b) and (2005c) as well as an earlier book we defined a quantum Turing machine, and quantum computer, formalism that can be related to the new *Operator Logic*.

1. What is a Statement?

*It is remarkable that up to the present day it [logic] has not been able to
make a single step forward, and thus, to all appearances,
it may be considered as complete and finished.
Critique of Pure Reason (Second Edition) - Kant*

With all respect to Kant, Logic has progressed remarkably in the centuries since his comment. And the progress will continue since Logic is an experimental science, in our view, just as Physics. Our experience of the environment, our senses, our experiments, and our development of scientific and mathematical theories to account for these phenomena lead us to expand linguistically and intellectually – thus feeding the growth of logic constructs in the human mind.

This book will introduce a new formalism for Logic based on the operator formalism of Quantum Measurement Theory. A remarkable correspondence between Logic and Quantum Theory will unfold that will extend Logic via this new operator formulation. This formulation enables the standard Logical Paradoxes and Gödel's Undecidability Theorem to be understood and resolved.[1]

1.1 Types of Statements

Statements (or sentences[2]) may be considered as entities in themselves. They can be logically combined with other sentences to produce compound sentences using connectives such as *and, or, not, if ... then ...*, and so on. The set of sentences, and their combinations, were called the Sentential Calculus by Hilbert (1928). Statements generally have the property of being true or false although "nonsense" statements

---

[1] Gödel recognized the relation between logical paradoxes and his Undecidability Theorem. See Gödel (1992) p. 40.
[2] We use these words interchangeably.

are possible such as "Love is green", which have the property of being neither true nor false because of their meaninglessness.[3]

There is a deeper form of "nonsense" statements that are called logical paradoxes because they have an internal inconsistency that renders them neither true nor false. A classic example of a logical paradox that dates back to the fourth century BC is the Liar Paradox. The philosopher Eubulides[4] seems to have first stated the equivalent of the paradoxical sentence: "The statement that I am now making is a lie." Clearly if it is true, then it is false; if it is false then it is true. Thus it is inherently paradoxical.

We shall not consider the sentential calculus in more detail at this point because it will follow automatically from the Operator Logic formalism that we will develop later.

Rather we will start by considering the internal logical form of statements and develop the *first order predicate calculus*. The first order predicate calculus deals with simple statements of the form (in English):

subject predicate

as well as more complex forms. Predicates typically contain a verb and often express a property. For example, a simple predicate is "is mortal". A statement generated from this predicate would have the form:

X is mortal.

where X is a subject such as Socrates. In more complex cases there can be multiple subjects and multiple clauses with differing predicates in a sentence.

We will next list some of the better known logical paradoxes.[5] Then we will formulate Operator Logic for statements and sequences of statements with a view towards resolving the paradoxes.

---

[3] Shakespeare could have given a metaphorical meaning to this sentence. But as a scientific, psychological statement it is meaningless.
[4] According to legend, Eubulides was killed by his fellow citizens for creating this logical paradox. Logic was a serious issue in those times.
[5] Needless to say these paradoxes each represent an infinity of variations. So the number of logical paradoxes is boundless.

1.2 The Logical Paradoxes (Antinomies) and Gödel's Undecidability Theorem

There is a traditional set of logical paradoxes. Most appeared in the 19th and early 20th centuries although some (such as the Liar paradox) date back to classical Greek times. We will briefly describe these paradoxes and then point out that Gödel's Undecidability Theorem has the same character (as Gödel himself pointed out in his celebrated paper).

Liar Paradox

The Liar Paradox as stated by Eubulides is: "The statement that I am now making is a lie." The simplest form of the paradox is a person saying, "I am lying." An impersonal form is "This sentence is false." They all have the character that if they are true, they are false. And if they are false, then they are true. Clearly they exhibit a form of self-contradiction.

Grelling Paradox

Some adjectives have the property that they apply to themselves. Some examples are "English", "short", "polysyllabic", and "kurz" which means "short" in German. Such adjectives can be called autological. Other adjectives do not apply to themselves. For example, "long" is not long. These adjectives can be called heterological. Is "heterological" heterological? If so, then it is autological. If not, then it is heterological. Both possibilities are self-contradictory. This paradox:

"Heterological is heterological."

shows how a word can, in itself, lead to a paradox.

Barber Pseudoparadox

The Town mayor issues an order: "The one village barber must shave those men in the village who do not shave themselves." If the barber shaves himself, he violates the order. If the barber does not shave himself he violates the order. A paradoxical situation! However there is a way out. The paradoxical situation is resolved if the tacit assumption that the barber is male is not true. A female barber or robot barber evades the paradox. Thus we have an example of an equivocal paradoxical situation that is resolved by a revised assumption: the barber is not male.

Berry Paradox

The number of positive integers that can be named in English in less than a fixed number of syllables is finite. Thus there must be a least integer that cannot be so named. However, " the least integer that cannot be named in English in less than fifty syllables" is an English name of less than fifty syllables. Thus the least integer has a name contrary to the assumption and thus the paradox.

Russell's Paradox

From experience we know we can consider classes of things; classes of integer numbers, classes of cars, and so on. We can consider classes of classes such as the class of all classes of cars in the various big cities of America. There are two interesting varieties of classes: proper classes and improper classes. *Proper classes* are classes, which are not members of themselves. For example, the class of all cars in China is proper. *Improper classes* are classes, which are members of themselves. For example, the class of all classes is a member of itself and thus improper.

Let us define the Russell class R as the class of all proper classes. Paradox: if R is a proper class, it is a member of itself, and is thus by definition an improper class. If R is an improper class, it is not a member of itself and therefore, by definition, it is a proper class. Thus there is no happy resolution of this paradox according to conventional logic.

Cantor Paradox

According to the theory of cardinal (infinite) numbers the set of all subsets of a set C has a higher cardinal number than C. If C is the set of all sets, then the preceding statement is a contradiction.

Burali-Forti Paradox

The Burali-Forti paradox is somewhat technical being based on the theory of transfinite (infinite) numbers. The theory proves a) every well-ordered set has a unique ordinal number; b) any set of ordinals, that is placed in a natural order such that each element contains all its predecessors, has an ordinal number which is greater than any preceding element in the set; and c) the set A of all ordinals placed in natural order is well-ordered. Then by theorems a and c, A has an ordinal number n. Since n is in A we see n < n by theorem b, thus establishing a contradiction.

Richard Paradox

The Richard paradox is concerned with the proposition: the set of all numerical functions is not enumerable. A commonly used argument to prove this proposition is the following. Suppose an enumeration existed symbolized by $f_n(m)$, which represents the nth function with argument m. Consider the function g defined by

$$g(n) = f_n(n) + 1$$

for any value of n. Let n_0 be the index number of g(n) in the enumeration:

$$g(n) = f_{n_0}(n) + 1$$

Then

$$f_{n_0}(n_0) = g(n_0) = f_{n_0}(n_0) + 1$$

which is a contradiction. Thus the set of all numerical functions is not enumerable.

Contrarian argument: Consider the set of all definable functions. Definable is taken to mean definable in some specific language with a fixed dictionary and grammar. Since the number of words in the language is finite, then the number of expressions is enumerable. Thus the set of expressions that form the definitions of definable functions is enumerable. Thus the set of definable functions is enumerable. Since the set of numerical functions is a subset of the set of definable functions it also must be enumerable. Thus the set of all numerical functions is enumerable.

The result of the preceding two arguments is a contradiction (paradox).

Gödel's Undecidability Theorem

Gödel's Undecidability Theorem, which has been the source of acclaim for many year's is of the same sort as the preceding paradoxes as Gödel himself stated[6]

> "The analogy between this result and Richard's antinomy[7] leaps to the eye; there is also a close relationship with the 'Liar' antinomy,[8] since the undecidable proposition [R(q); q] states precisely that q belongs to K, i.e. according to (1)[9], that [R[(q); q] is not provable. We are therefore confronted with a proposition which asserts its own unprovability."

Thus a successful resolution of these paradoxes suggests a successful resolution exists of the dilemma created by the Undecidability Theorem. We consider this resolution in detail in our discussion of the implications of Operator Logic.

1.3 The Many Aspects of a Statement

Although Logic has been a subject for study for 2500 years our understanding of its simplest unit, the sentence, is still a subject of controversy and, based on modern developments, gives us the uneasy feeling that we have somehow missed important aspects of the nature of a sentence.

Traditional View of a Statement

For more than two thousand years a sentence (statement) was a string of words arranged in a language dependent order. A sentence was either true or false. It had parts (a subject and predicate) but the emphasis was on the combination of sentences and their use in syllogisms and proofs.

---

[6] Gödel (1992) p. 40.

[7] Antinomy is a synonym for paradox.

[8] Braithwaite footnotes this statement with, "Every epistemological antinomy can likewise be used for a similar undecidability proof."

[9] Eq. (1) of Gödel is n ε K ≡ $\overline{\text{B e W}}$ [R(n); n] where BeW x means x is provable and the bar over B e W indicates negation.

The Fregian View of a Statement

The seminal logician Frege (and others) created a new view of a sentence. It was not merely a string of words that only had the property of being true or false. He introduced the concepts of the "*Bedeutung*" of a sentence and the "*sense*" of a sentence. Unfortunately Bedeutung means different things in different contexts.[10] The Bedeutung of a sentence is usually interpreted as its truth value.[11] However the manner in which Frege discusses Bedeutung suggests that it encompasses more than simply whether a sentence is true or false. For Frege states the parts of a sentence have a Bedeutung[12] upon which the Bedeutung of the sentence depends; and he further states that the Bedeutung of a sentence is also dependent on the context (set of sentences) in which it appears. Thus Bedeutung seems to mean "meaning" as well.

Bedeutung of a Sentence

A sentence's Bedeutung can be difficult to determine although it is either true or false in Frege's view. A sentence's Bedeutung is contextual. In view of these complexities one might expect that Frege had provided a general procedure to calculate the special case of the Bedeutung of a single sentence, and the more general case of the Bedeutung of a sentence within the context of a set of sentences.[13] Unfortunately, he did not for several reasons. Firstly, the Bedeutung of Bedeutung depends on the item: for a sentence the Bedeutung is true or false; the Bedeutung of a proper name is the object the name represents; relations and concepts have no Bedeutung; and so on.[14] Thus the inconsistency of the meaning of Bedeutung is one reason for the lack of a procedure to calculate the Bedeutung of a sentence or passage. Another reason for this omission is the difficulty inherent in the use of human languages, which can express a statement in many ways.

---

[10] In German Bedeutung is usually taken to mean "meaning".

[11] Frege (1997) p. 34.

[12] But the Bedeutung of the parts is variable and sometimes non-existent: the Bedeutung of a proper name is the object it represents. The phrase "least rapidly convergent series" does not have a Bedeutung since it doesn't exist. So the Bedeutung of a part of a sentence appears to be the object for which the part is the name.

[13] The Bedeutung of a set of sentences (a passage of text) is another issue that Frege apparently did not address to the author's knowledge.

[14] See "On Sinn and Bedeutung" in Frege (997).

Later it will be clear that the Bedeutung of a word, sentence, and set of sentences is implicit in the operators within the framework of Operator Logic.

The Sense of a Sentence

Frege also introduces another aspect of a sentence: the *sense* of a sentence. This concept is also somewhat vague and perhaps inconsistent.[15] In my view Frege's concept of sense is best understood to be semantic meaning. Tarski developed a concrete method of combining a sentence with its semantic meaning through the logical relation:

$$SM \text{ is true if, and only if, } S \qquad\qquad (1.5.1)$$

where S is a sentence and SM is its semantic meaning. For example,

"FeO_2 is rust (oxidized Fe)." is true if, and only if, "Rust is oxidized iron."

Thus the *sense* of the sentence "Rust is oxidized iron." is "FeO_2 is rust (oxidized Fe)." since it explains the meaning of the sentence. We thus identify the *sense* of a sentence with the semantic content of the sentence. "FeO_2 is rust (oxidized Fe)." is not the idea of the sentence (in the mind of a hypothetical reader) because the idea requires the mind to know what Fe and FeO_2 are.

The Idea of a Sentence

The sense of a sentence is the semantic content of the sentence as noted in the preceding section. However there is a difference between the sense of a sentence and the idea of a sentence as noted by Frege and others. We define the *idea* of a sentence as the interpretation of the sense of a sentence in the mind of the reader of the sentence. Different readers will read the semantic content of a sentence differently due to their

---

[15] See again "On Sinn and Bedeutung" in Frege (997) for examples of the vagueness of "sense." Compare the concept of sense; in Frege's astronomy example (p. 155) where the moon is the Bedeutung, the telescope image is the sense, and the astronomer's view of the telescope image corresponds to the idea; with Frege's discussion of the words "evening star" and "morning star" where he asserts the thoughts of these phrases are their sense (p. 156). Since thought and idea would seem to be the same thing in both these examples, his use of the word sense is not consistent.

differing education, differing experience, and their differing mechanisms of thought. The mechanism of thought of people varies. The variation is due partly to previous training and experience (The human brain can be "rewired" according to modern brain research.), and partly due to genetic differences. An extreme version of this point was the Platonic theory that education was relearning the knowledge we had in a prior reincarnation. The idea that a sentence invokes in a reader generally depends on the reader's mindset. There are of course simple statements such as "The sky is blue." that generate virtually the same idea in all readers. But the majority of statements in discourse tend to generate different ideas in readers. The idea of a sentence, being internal to the individual, will not be embodied in our Operator Logic. Indeed it is difficult to view it as a part of Logic – it is more properly viewed as a part of psychology or phenomenology.

Our View of a Statement

We will develop our Operator Logic with the above stated features of a statement in mind. Operator Logic will embody these features in a uniform way within the operator framework.

1.4 Deductive Systems and Their Calculi

There are many theories of various aspects of mathematics, physics, and other sciences. Some examples include Euclidean geometry, quantum field theory, elementary number theory and so on. All or parts of these theories can be formalized with a set of axioms expressed in terms of primitive concepts, from which theorems are derived. The axioms plus the set of derivable theorems constitutes an *axiomatic system.*[16]

Axiomatic systems can be viewed as consisting of two types: deductive systems[17] and calculi.[18] A *deductive system* has *semantic*[19] characteristics—it defines the primitive terms of the axioms (usually in terms of "real" things such as real lines, real physical particles and their properties, real properties of integers, and so on). Thus the implications

---

[16] Kleene (1967) pp. 198 – 201.
[17] Another name for some deductive systems is mathematical-deductive system.
[18] Other names used for calculi (singular calculus) are formal system, formalism, or logistic system.
[19] We use semantic in its usual sense as the "meaning" of something.

of the axioms are "real" as well and describe the reality associated with the deductive system. In physics a deductive system describes an aspect of nature. Examples include the Landau-Ginzberg theory of superconductivity, and Newton's mechanics. In mathematics examples are Euclid's geometry and Gödel's theory of integer numbers. Statements derived in a deductive system are true – a semantic property.

A calculus is an axiomatic system without a definition of primitive terms. Theorems are derived following specified rules of derivation with no reliance on semantics. Thus they have syntax without semantics. Most calculi have one or more corresponding deductive systems. It is possible to have a calculus without any known corresponding deductive system. Despite that, a calculus is best viewed as the skeleton of a deductive system. The deductive system constitutes the flesh.

The discussion of a *calculus*, its definition and its consequences, is the *metatheory* or *metalanguage* or *syntax language* of the calculus. The investigation of a calculus is done within the metalanguage and called *metamathematics* or *proof theory*. The calculus itself is often called the *object theory* or *object language*.

While there is much more to be said on this topic, the preceding discussion is all that we need to develop the features of Operator Logic. The reader is referred to the References at the end of this book, and other books on Logic, for more details.

1.5 The Truth Value of a Statement and the Resolution of Paradoxical Statements – Simplified First Look

Statements (sentences) are usually thought of as being either true or false although there are statements that can be characterized as *nonsense statements* that are neither true nor false because their content is usually fictional. We will exclude nonsense statements from our discussion.

We now consider the question of whether every statement is true or false based on an analysis of its constituent parts. We start with the paradoxical statement:[20]

$$\text{This sentence is false.} \qquad (1.7.1)$$

---

[20] The Liar paradox; also known as Epimenides' paradox.

If it is true, then it is false. If it is false then it is true. Thus the paradox. Clearly, this statement is but one of an infinity of possible paradoxical statements (see earlier discussion) of a similar character although many statements of this infinity would require a deeper analysis to uncover its paradoxical nature.

Although numerous proposals have been made to eliminate paradoxical statements that would represent a flaw in the development of a logical system, none of them have been totally convincing. Some attempt to postulate such statements away. Some sweep them under the rug where they generate problems in other areas. So the problem of paradoxical statements remains.

A generalization of the constituent properties of a statement is needed to conclusively eliminate paradoxical statements from Logic. This generalization has an analogue in the measurement theory of Quantum Mechanics, which we describe in the following chapter.[21] The generalization is based on the axioms:

1. Every statement appears in a universe of discourse. A *universe of discourse* is a set of statements and phrases. We denote a universe of discourse state in the form $|\Omega>$ and treat it as a state in a Hilbert space using the notation and concept of quantum state introduced by Dirac (described later).

2. Each statement, and each part of a statement, in a universe of discourse has an associated logical status operator S, which is a projection. These operators when applied to the universe of discourse state $|\Omega>$ have two possible types of results. They can yield a zero value or a non-zero state. (A statement yielding a zero value does not have a truth value. A statement yielding a non-zero value does have a truth value – it is either true or false.) Thus $S|\Omega> = 0$ indicates undecidable, while the expectation value (inner product) $<\Omega|S|\Omega> \neq 0$ indicates decidable – either true or false.

3. The status operator S of a statement is the product of the status operators of the phrases (terms) in the statement. A phrase can be

---

[21] The reader who may be apprehensive of the use of Hilbert spaces, which is implicit in this introductory discussion, may be reassured somewhat by noting that Hilbert spaces together with their mathematical rules can be viewed as a language on a par with human languages and symbolic languages for use in Logic. We address this issue in more detail in a subsequent chapter.

one word or a finite set of words comprising a concept (an idea) or object. So symbolically we can write

$$S = S_1 \, S_2 \, S_3 \, S_4 \, S_5 \, \ldots \qquad (1.7.2)$$

where S_1, S_2, S_3, S_4, S_5 and so on represent the status operators of the phrases in the statement S.

4. The status operator of a phrase S_i is determined by its use in the universe of discourse. It is a projection operator satisfying $S_i^2 = S_i$.
5. Projection operators are not necessarily commutative. (This will be discussed in more detail subsequently.)

If we now consider the earlier paradoxical statement 1.7.1 and form a universe of discourse by adding

$$\text{This sentence is true.} \qquad (1.7.3)$$

we find the universe of discourse consists of the subject "This sentence" and the predicates "is true" and "is false." We now assign status operators:

Item	Status Operator
"This sentence"	P
"is true"	P
"is false"	$1 - P$

$$(1.7.4a)$$

with the result

Statement	Status Operator Product
"This sentence is true"	$PP = P$
"This sentence is false"	$P(1 - P) \equiv 0$

$$(1.7.4b)$$

where P is a projection operator. Note $P(1 - P) = (P - P) = 0$. Consequently we find 1.7.1 does not have a truth value while eq. 1.7.3 is does have a truth value. A paradoxical statement becomes a non-paradox by the introduction of a new feature for statements, subjects, and predicates: namely status operators.

Another simple example is the Grelling paradox:

"Heterological is heterological."

If we define a simple universe of discourse with the status operators:

Item	Status Operator
"Heterological"	P
"is heterological"	$1 - P$

$$(1.7.5a)$$

then

Statement	Status Operator Product
"Heterological is heterological."	$P(1 - P) \equiv 0$

$$(1.7.5b)$$

we see the paradox is resolved: the statement does not have a truth value. A predicate containing a word, which is also a subject, may have a different status operator.

In the case of more complex universes of discourse, one introduces as many status operators as needed: P_1, P_2, P_3, and so on in such a way as to exclude paradoxical statements. This can be performed systematically (in principle) by establishing all combinations of subjects and predicates that are paradoxical and defining projection operators so as to exclude paradoxical statements from the set of valid statements.

The discussion in this section is meant to give the flavor of Operator Logic without obscuring it with numerous technical details and without the introduction of Quantum Measurement Theory (the general basis of Quantum Physics), to which it is analogous.

2. Quantum Measurement Theory

2.1 What Does Quantum Measurement Theory Have to Do With Logic?

Quantum Measurement Theory is an abstract mathematical formalism that was invented by P. A. M. Dirac[22] in 1931 to describe the fundamental nature of Quantum Theory based on the analysis of physical measurements[23] as "filtrations." An experiment proceeds in a series of stages that "filter" or transform the physical system under study until the end of the experiment is reached. Each stage of filtering is mathematically represented by a projection operator that only allows the desired part of the experimental state to pass to the next stage. The complete experiment is represented by the product of the projection operators of each stage of the experiment.

A statement also has a series of "stages" that we recognize as the consecutive words or phrases, or subjects or predicates.[24] In the preceding chapter we saw examples of how one could eliminate paradoxical statements by associating a status operator with each part of a statement. In a manner similar to Dirac's representation of an experiment as a product of projection operators, we saw the "value" of a statement as the product of the status operators of its parts.

Thus there is a correspondence between the quantum measurement theory representation of an experiment and the product of status operators that determine whether a statement has a truth value. This correspondence will be explored in detail in chapter 3. We will also see in chapter 7 that quantum measurement theory leads us to consider

---

[22] See Dirac (1931) for an exposition of his theory.

[23] Quantum Measurement Theory is more general than Quantum Mechanics. It also applies to the successors of Quantum Mechanics: relativistic Quantum Mechanics and Quantum Field Theory—which were developed in the period from the mid-1930's to 1960 (and is still under investigation). *Thus one can say that Quantum Measurement Theory captures the deepest understanding we have of the nature of physical phenomena.*

[24] The order of the stages is language dependent but the order is always linear.

statements of the quantum probabilistic sort giving us Quantum Operator Logic.

The fact that we can set up a detailed correspondence between Quantum Measurement Theory and the logic of statements is of great importance. Quantum Measurement Theory is at the deepest level of Physics that we know of. Nature cannot be illogical, inconsistent or undecidable.[25] *Therefore the Operator Logic that we have created, which is based on an analogy with Quantum Measurement Theory, also cannot be illogical, inconsistent or undecidable. Thus Operator Logic is guaranteed to be "correct."* One cannot say this of other formulations of Logic because, in the final analysis, they are based on intuition, conjecture, and inspired guessing.

We see Nature as providing the ultimate guide to the fundamental nature of Logic. And we view Hilbert spaces of operators as the ultimate language of Logic. Human and symbolic languages created by Mankind are flawed because they inevitably lead to paradoxes—as many logicians have remarked. The fact that Hilbert spaces are a complex construct does not preclude their use as the fundamental language of Logic – because Hilbert space features are well known and well defined.

2.2 Experiments as Statements

As we noted earlier Physics experiments, both classical physics and quantum physics experiments, can be viewed as statements. The statements are often of the modus ponens sort (if … , then …) but sometimes they are assertions. For example one might perform an experiment to measure the speed of light. The experiment should show that the speed of light is approximately 186,000 miles per second and, therefore, can be said to embody the assertion, "The speed of light is approximately 186,000 miles per second." An example of an experiment of the modus ponens type would be to build a ruby laser and measure the frequency of the light it produces. The statement corresponding to this experiment is "If I create a ruby laser and run it, then laser light of frequency nnn will be produced."

---

[25] This statement includes quantum effects. They are probabilistic in nature; yet their probabilities are calculable in quantum theory.

Thus every classical physics experiment corresponds to a statement. In the case of experiments that are quantum mechanical in nature the corresponding statements are quantum probabilistic in nature.

For example, if I perform the experiment of colliding two electrons "head on" at low energy so no new particles are produced and if I detect the percentage of electrons after the collision at angles between 60 degrees and 90 degrees with respect to the original direction of the electrons, then I can only make a probabilistic statement such as: "If I collide two electron at a center of mass energy of 20 electron volts, then mm% of the emergent electrons will be traveling at an angle between 60 degrees and 90 degrees with respect to the original direction of motion." This modus ponens statement is based on quantum mechanics and therefore is quantum probabilistic in nature.

Another way to phrase it in two statements that brings out the quantum probabilistic aspect more clearly is "If I collide two electron at a center of mass energy of 20 electron volts, then the emergent electrons will be traveling at an angle between 0 degrees and 60 degrees with respect to the original direction of motion with 80% probability."[26] and "If I collide two electron at a center of mass energy of 20 electron volts, then the emergent electrons will be traveling at an angle between 60 degrees and 90 degrees with respect to the original direction of motion with 20% probability." Thus we can only make statements of a probabilistic nature in quantum physics.

2.3 Quantum Measurement Theory

Quantum Measurement Theory[27] is built around the notion of operators that represent physically measurable quantities such as the position of a particle, the momentum of a particle, the spin of a particle, and so on. Projection operators appear in measurement theory to project (select) specific values of physically measurable properties at various

---

[26] The probabilities cited are simply for illustrative purposes and are not the actual calculated probabilities in quantum theory.

[27] This section is based on Dirac original work (1931) – the originator of Quantum Measurement Theory. See Gottfried (1989), Messiah (1965), and Mackey (1963) as well as certain papers of Schwinger. The author was struck by the content of Gottfried's chapter on Dirac's Transformation theory some forty years ago (1969) while a graduate student and felt there was a further depth to Dirac's work that wanted bringing forth. This depth surfaced in the creation of the present work in 2008 after this author developed an axiomatic derivation of the Standard Model and after a study of the progress of Logic in the past 150 years.

stages during the course of an experiment. Thus *there are two types of operators in measurement theory: 1) those operators whose eigenvalues are the values of physically measurable properties, and 2) those operators which project*[28] *(select, filter) the value of a physically measurable property at some stage of an experiment.* We will call the operators of the first type *eigenvalue operators,* and the operators of the second kind *filter or measurement (or status) operators.*

An example of an eigenvalue operator is the momentum operator of a particle – usually denoted p. If a particle has momentum p_0 it has a corresponding quantum state that is often denoted $|p_0>$ using Dirac's bra-ket notation. Then applying the momentum operator to the state yields the momentum p_0 of the particle as its eigenvalue:

$$p| p_0> = p_0|p_0> \qquad (2.3.1)$$

An example of a filter (measurement) operator is a projection operator that selects particles of a specific momentum. An experimental apparatus can have a stage that only allows particles of a specific momentum to proceed to the next stage of the experiment. So we can define a corresponding filter (projection) operator $P(p_i)$ that only allows particles of momentum p_i through. As a result

$$P(p_1)|p_0> = 0 \qquad (2.3.2)$$

while

$$P(p_1)|p_1> = |p_1> \qquad (2.3.3)$$

If we apply the projection operator $P(p_1)$ to a state consisting of a particle that partly has momentum p_0 and partly has momentum p_1 then only the p_1 part of the state proceeds to the next part of the experiment:

$$P(p_1)(a|p_1> + b|p_0>) = a|p_1> \qquad (2.3.4)$$

where a and b are constant weight factors (actually the "square roots" of the quantum probabilities that the particle is in the momentum states $|p_1>$ or $|p_0>$). Note that only the $|p_1>$ state appears on the right side of eq. 2.3.4 and continues to the next stage of the experiment.

---

[28] The selection process is called filtering. Particle states with an eigenvalue different from the specified eigenvalue are "prevented" from going to the next stage in an experiment.

Let us now assume we have some quantum system that has an associated set of physically measurable quantities[29] such as momentum and so on. Each measurable quantity has a corresponding eigenvalue operator that we denote as A_i (where $i = 1, 2, 3, ...$) whose range of eigenvalues are the possible values of measurements of that quantity. We denote the set of eigenvalues of A_i as $\{a_{ij}\}$ where j specifies the jth eigenvalue of A_i. The set of eigenvalues of an eigenvalue operator is only restricted by the requirement that the eigenvalues are numeric. The set can be discrete, infinite or finite in number, or continuous.

At any stage of an experiment the system can be filtered to "select" the "parts" of the system that have certain eigenvalues for certain observables. We represent this process mathematically with measurement (filter) operators[30] that select specified eigenvalues. In the case where we filter only one observable we denote[31] the operator as $M(A_i(a_{ij}))$ where a_{ij} is the jth eigenvalue of the ith observable A_i. For example, if we filter the initial state of a system $|\Omega_0>$ to place it in the state $|\Omega_1>$ where the 2nd observable's eigenvalue is a_{21} then we represent this with the expression:

$$|\Omega_1> = M(A_2(a_{21}))|\Omega_0> \qquad (2.3.5)$$

Measurement operators are projection operators since if we filter the system twice for the same eigenvalue the system remains the same as it was after the first filtration.

$$|\Omega_1> = M(A_2(a_{21}))|\Omega_1> = M(A_2(a_{21}))M(A_2(a_{21}))|\Omega_0> \qquad (2.3.6)$$

Thus

$$M(A_i(a_{ij})) = M(A_i(a_{ij}))M(A_i(a_{ij})) \equiv M(A_i(a_{ij}))^2 \qquad (2.3.7)$$

If we filter the system twice, but for different eigenvalues of an operator, we find that the result is zero – nothing goes to the next stage of the experiment. Thus

---

[29] These quantities are called observables in physics.

[30] As Mackey (1963) points out: every hermitean operator in a Hilbert space has an associated measurement (filter) operator that is a projection. Eigenvalue operators are hermitean operators.

[31] We use 'M' as the obvious general label (name) of Quantum <u>M</u>easurement Operators.

$$M(A_i(a_{ij}))M(A_i(a_{ik})) = 0 \tag{2.3.8}$$

if $j \neq k$. Eqs. 2.3.7 and 2.3.8 can be combined using a form of the Kronecker delta function

$$M(A_i(a_{ij}))M(A_i(a_{ik})) = \delta(a_{ij}, a_{ik})M(A_i(a_{ij})) \tag{2.3.9}$$

where

$$\delta(a_{ij}, a_{ik}) = 0 \quad \text{if } a_{ij} \neq a_{ik}$$
$$= 1 \quad \text{if } a_{ij} = a_{ik}$$

where a_{ij} and a_{ik} are eigenvalues of an observable A_i of the experiment. The quantity 0 (zero), indicates all states are filtered out by the consecutive measurement filters if $a_{ij} \neq a_{ik}$:

$$M(A_i(a_{ij}))M(A_i(a_{ik})) |\Omega_0\rangle = 0 \tag{2.3.10}$$

If the expression $\delta(a_{ij}, a_{ik}) = 1$ (one) in eq. 2.3.9, then it means that the repeated consecutive measurement filter does not do any additional filtering.

A simple algebra of measurement operators can be developed based on physical grounds:

1) Consider a stage in an experiment where we wish to let particles with momentum p_0 and p_1 through but wish to exclude all particles of other momentum. Then it is clear that the corresponding projection operator is

$$P(p_0) + P(p_1) \equiv P(p_1) + P(p_0) \tag{2.3.11}$$

Note the commutativity of the addition of measurement operators.

2) If we wish to let particles with any of three specified momenta (p_0, p_1, and p_2) through then we see the corresponding projection operator is

$$(P(p_0) + P(p_1)) + P(p_2) \equiv P(p_0) + (P(p_1) + P(p_2)) \equiv P(p_0) + P(p_1) + P(p_2) \tag{2.3.12}$$

The addition of measurement operators is associative.

3) If at some stage of an experiment we wish to pass particles of any momentum through (in other words, no filtering) then there is an "identity filter" I satisfying

$$\sum_{k} P(p_k) \equiv I \qquad (2.3.13a)$$

where the sum is over all momenta p_k. (Generally the set of momenta form a continuous range of values but this is not of importance at this point in our development.) The symbol I represents the identity filter (measurement) that lets everything through to the next stage of an experiment.

More generally

$$\sum_{k} M(A_i(a_{ik})) = I \qquad (2.3.13b)$$

and

$$I|\Omega_0> = |\Omega_0> \qquad (2.3.13c)$$

The measurement operator that does not let anything through to the next stage is denoted 0 (zero).

Up to this point in this section we have been considering one observable, and its corresponding eigenvalue operator and measurement operator. Now we consider a set of observables of some system and their corresponding operators. (The set may or may not constitute the set of all observables of the system.) There are two general cases of interest: the case where all the eigenvalue operators commute with each other:

$$[A_i, A_j] = 0 \qquad (2.3.14)$$

for all i and j; and the case where some of the operators do not commute with each other:

$$[A_i, A_j] \neq 0 \qquad (2.3.15)$$

for some i and j.

A set of operators that all commute with each other is called a *compatible set.* A set of operators, some of which do not commute with each other, is called a *incompatible set.* (See chapter 7.)

These cases can be reduced to considering a pair of operators that commute, and a pair of operators that do not commute. Let us consider the commuting pair of operators A and B with the eigenvalue equations $A|\Omega> = a|\Omega>$ and $B|\Omega> = b|\Omega>$ for the $|\Omega>$ state of some system. Since A and B commute there exist simultaneous eigenstates of both operators, which we denote $|ab\Omega'>$ that satisfy

$$A|ab\Omega'> = a|ab\Omega'> \quad \text{and} \quad B|ab\Omega'> = b|ab\Omega'> \quad (2.3.16)$$

The corresponding projection operators $P(A(a))$ and $P(B(b))$ filter the state of the system. Because of eq. 2.3.16 it is clear that the projection operators (filters) for A and B must also commute:

$$[P(A(a)), P(B(b))] = 0 \quad (2.3.17)$$

In words, the order of filtering (or measurement) is irrelevant if the corresponding eigenvalue operators commute.

For a set of compatible (commuting) eigenvalue operators that corresponds to the complete set of observables of a system we can define a measurement operator

$$M(a_{1j}, a_{2k}, a_{3m}, \ldots) = \prod_i M(A_i(a_{ik_i})) \quad (2.3.18)$$

that filters the system assigning a specific eigenvalue for each operator in the set of compatible operators. The measurement operators in this product all commute with each other.

If two eigenvalue operators do not commute, then it is not possible to have a state that is simultaneously an eigenvector of both operators. In this case the order of filtration does matter – the corresponding measurement operators do not commute. We discuss sets of incompatible eigenvalue operators in chapter 7.

In general we can define the multiplication of measurement operators by

$$M(A_i(a_{ik}))M(A_j(a_{jm})) \quad (2.3.19)$$

which means apply the measurement operator $M(A_j(a_{jm}))$ first and follow it with the application of $M(A_i(a_{ik}))$ next (read right to left). As noted above these measurement operators commute if, and only if, A_i and A_j commute.

Measurement operator multiplication is associative:

$$(M(A_1(a_1))M(A_2(a_2)))M(A_3(a_3)) \equiv M(A_1(a_1))(M(A_2(a_2))M(A_3(a_3)))$$

$$(2.3.20)$$

At this point we suspend our discussion of Quantum Measurement Theory temporarily until chapter 7. The remaining topics of Quantum Measurement Theory bring in probabilities and are thus only relevant for Quantum Probabilistic Operator Logic.[32]

Chapter 3 is concerned with absolute (no probabilities) statements that are true or false or have no logical value: Classical Operator Logic.

---

[32] Implicit in our discussion of Quantum Measurement Theory was the role of the Observer – the "individual" that sets up an experiment, performs filtering, and measures the values of observables during, and after, an experiment. We discuss Observers in Appendix A – particularly in relation to the Platonic connection of Operator Logic and Reality.

3. Classical Operator Logic

*The purpose of the symbolic language in mathematical logic
is to achieve in logic what it has achieved in mathematics,
namely, an exact scientific treatment of its subject matter.
Principles of Mathematical Logic - D. Hilbert & W. Ackermann*

3.1 Operator Representation of a Statement

We have discussed the correspondence between statements and physical experiments in previous chapters. In chapter 1 we introduced status operators, which are projection operators just like the Quantum Measurement operators that we introduced in chapter 2. It is clear that the analogy goes deeper if we interpret statements as the "eigenvalues" of sequences of eigenvalue operators where the "eigenvalues" are words, phrases, or sequences of logical symbols.[33] We will call these strings of characters "eigenvalues" as well, although eigenvalues are normally numeric.[34] A *term* is one or more words or symbols. The terms that compose a statement can be viewed as occurring in consecutive "stages" (although the exact ordering is language dependent.)

We will develop an operator formalism that will yield statements as a sequence of eigenvalues of operators. This formalism is not quantum probabilistic at this point. (The quantum generalization will be presented in chapter 7.)

A simple example that illustrates the concept is to define a state vector with two "eigenvalues" such as

$$|\Omega_1> = |\text{"The car", "is red"}> \qquad (3.1.1)$$

and two operators, Subject and Predicate, satisfying the "eigenvalue" equations:

---

[33] Thus terms can be primitive terms, defined terms, functions, variables, and logical and mathematical symbols.

[34] We convert terms to numeric values – Gödel numbers – at a later point in the discussion. So treating terms as eigenvalues is legitimatized by a mapping from terms to numbers.

$$\text{Subject}|\Omega_1\rangle = \text{"The car"} |\Omega_1\rangle \qquad (3.1.2)$$
$$\text{Predicate}|\Omega_1\rangle = \text{"is red"} |\Omega_1\rangle \qquad (3.1.3)$$

In the present example the "eigenvalues" are strings of characters. Mathematically, we can relate each string of characters (symbols) in a one-to-one manner[35] to a real numeric value by mapping strings into Gödel numbers, denoted gn(i). By mapping strings into Gödel numbers[36] we can establish numbers as eigenvalues for our eigenvalue operators. The set of operators is then defined within the context of a Hilbert space formalism. Thus we reduce the various eigenvalue operators of interest to self-adjoint Hilbert space operators.

3.1.1 Gödel Numbers

Gödel numbers are real positive numbers. They can be viewed as the eigenvalues of self-adjoint operators A_i corresponding to Subject and Predicate in the above:

$$A_{\text{Subject}}|\Omega_1\rangle = gn(\text{"The car"})|\Omega_1\rangle \qquad (3.1.4)$$
$$A_{\text{Predicate}}|\Omega_1\rangle = gn(\text{"is red"})|\Omega_1\rangle$$

and $|\Omega_1\rangle$ can be viewed as the direct product of eigenstates of A_{Subject} and $A_{\text{Predicate}}$

$$|\Omega_1\rangle = | \, gn(\text{"The car"})\rangle| \, gn(\text{"is red"})\rangle \qquad (3.1.5)$$

Thus using plus signs as separators we express the statement as

$$(A_{\text{Subject}} + A_{\text{Predicate}})|\Omega_1\rangle = \text{"The car is red"}|\Omega_1\rangle \qquad (3.1.6)$$

A string expression such as

$$\text{AyyBx}$$

is assigned a Gödel number by combining the assigned numerical values of each symbol in the string. The numerical values become exponents of

---

[35] Due to the "Fundamental Theorem of Arithmetic", namely that every integer number greater than one, which is not a prime itself, has a unique representation in terms of prime number factors.

[36] It should be noted that Leibniz developed a procedure to map strings to integers. See Rescher (1967) p. 132 and cited references to Leibniz.

a product of prime numbers greater than one. For example the symbols in the preceding string expression could have the symbol numbers: $A \equiv 3$, $y \equiv 9$, $B \equiv 5$ and $x \equiv 7$. Then its G̲ödel n̲umber is

$$gn(\text{"AyyBx"}) = 2^3 3^9 5^9 7^5 11^7 \qquad (3.1.7)$$

where the numbers being exponentiated are the ordered set of prime numbers.

In general, if E is an expression (string) of symbols, β_1, β_2, β_3, β_4, ... β_n and v_1, v_2, v_3, v_4, ... v_n is the sequence of integers corresponding to these symbols, then the Gödel number of E is the integer

$$gn(E) = \prod_{m=1}^{n} \Pr(m)^{v_m} \qquad (3.1.8)$$

where $\Pr(m)^{v_m}$ is the mth prime number raised to the power v_m with the 1st prime number taken to be 2. If E is an empty string (containing no symbols) then $gn(E) = 1$.

The definition of Gödel numbers, and their consequent uniqueness, results in the following conclusions and definitions:

1. If $g = gn(E)$, then the inverse relation is defined to be $E = gi(g)$.

2. If E_1 and E_2 are expressions (terms) such that $gn(E_1) = gn(E_2)$, then $E_1 = E_2$.

3. If E_1, E_2, E_3, ... , E_n are a finite sequence of expressions, then the Gödel number of the sequence is

$$gn(E_1, E_2, E_3, ... , E_n) = \prod_{m=1}^{n} \Pr(m)^{gn(E_m)} \qquad (3.1.9)$$

4. The Gödel number of an expression is never equal to the Gödel number of a sequence.

5. If $E_1, E_2, E_3, \ldots, E_n$ and $E'_1, E'_2, E'_3, \ldots, E'_n$ are both finite sequences of expressions and

$$gn(E_1, E_2, E_3, \ldots, E_n) = gn(E'_1, E'_2, E'_3, \ldots, E'_m) \quad (3.1.10)$$

then $n = m$, and for $k = 1, 2, \ldots, n$

$$E_k = E'_k$$

and the strings to which they correspond are identical.

3.1.2 Statements Expressed in Gödel Numbers

Statements can be expressed in terms of Gödel numbers (eqs. 3.2.3 and 3.2.9 below). In this subsection we point out the general interpretation of expressions involving eigenvalue operators.

If eigenvalue operators are added as in eq. 3.1.6 then the sum of eigenvalue operators can be viewed as a symbolic list of Gödel numbers that is not multiplied. As we read a statement from word to word the statement "unfolds" and becomes particularized, more precise and more specific. In this sense the symbolic sum of eigenvalue operators denotes the statement when applied to the state that represents the system.

Placing eigenvalue operators adjacent to each other[37] is interpreted as equivalent to the logical relation "or." This can be seen in an extension of the simple example considered earlier. We add an additional term "is green" with operator $A_{\text{is green}}$ to the example and use the state

$$|\Omega_2\rangle = |\, gn(\text{"The car"})\rangle|\, gn(\text{"is red"})\rangle|\, gn(\text{"is green"})\rangle$$

Then the statement

$$(A_{\text{Subject}} + A_{\text{is red}}A_{\text{is green}})|\Omega_2\rangle \equiv \text{"The car is red or is green."}|\Omega_2\rangle$$

exemplifies the "or" case. Alternately we could define "or" as a term with a corresponding Gödel number.

---

[37] This corresponds with the "or" notation of Hilbert (1950) p. 12 where $XY = X \lor Y$.

$(A_{Subject} + A_{is\ red} + A_{or} + A_{is\ green})|\Omega_3> \equiv$ "The car is red or is green."$|\Omega_3>$

where

$$|\Omega_3> = | \text{gn("The car")}>| \text{gn("is red")}>| \text{gn("is green")}>| \text{gn("or")}>$$

The "or" operation for statements corresponds to adding quantum measurement operators together. It is a filtration stage where states, which are a composite of two states with different eigenvalues, are allowed to go to the next stage of an experiment.

3.2 Hilbert Space for a Universe of Discourse

We will define two types of universes of discourse: a semantic universe of discourse and a calculus universe of discourse. A *calculus universe of discourse* consists of undefined primitive terms, axioms, and theorems derived from the axioms. Statements in a calculus are provable, not provable or undecidable.

A *semantic universe of discourse* consists of primitive terms defined in some fashion, axioms, and theorems derived from the axioms. Statements in a semantic universe are true, false, or undecidable.

A calculus universe of discourse can be viewed as a semantic universe of discourse stripped of its definition of primitive terms and reduced to a "skeleton of algebra" with truth reduced to provability.

3.2.1 The Semantic Universe of Discourse

A *semantic universe of discourse* consists of a set of primitive and defined terms, related to some reality,[38] that are used to formulate the axioms and theorems of a deductive system. The terms are mapped to Gödel numbers. Each term E has a unique Gödel number gn(E). Each Gödel number is assumed to be an eigenvalue of a unique, self-adjoint, eigenvalue operator that we denote A_E which has two eigenvalues gn(E) and $0 = $ gn(""). To each of these eigenvalues there corresponds eigenstates denoted $|E>$ and $|0>$ respectively. If there are n primitive terms then a state of the universe of discourse is a product of eigenstates with the form:

---

[38] Related to Reality in the sense that each primitive term is "defined" in terms of real world features.

$$|E_1, E_2, E_3, \ldots, E_n> = \prod_{m=1}^{n} |E_m> \tag{3.2.1}$$

where each E_i can be a string or its Gödel number or zero. Note: since n is the total number of terms in the semantic universe of discourse it must be finite since the number of words in the English language, or any language expressing semantic meanings, is finite. Each eigenvalue operator A_{E_k} satifies the equation

$$A_{E_k}|E_1, E_2, E_3, \ldots, E_n> = gn(E_k)|E_1, E_2, E_3, \ldots, E_n> \tag{3.2.2}$$

for k ε [1, n]. *All eigenvalue operators in a universe of discourse in this chapter commute and are thus compatible.*
    A semantic statement in this notation has the symbolic form

$$\sum_{n \in \{m\}} A_{E_n} \tag{3.2.3}$$

If applied to the eigenstate eq. 3.2.1 we obtain the symbolic sum

$$\sum_{n \in \{m\}} A_{E_n}|E_1, E_2, E_3, \ldots, E_n> = \sum_{n \in \{m\}} gn(E_n)|E_1, E_2, E_3, \ldots, E_n> \tag{3.2.3a}$$

where {m} is the set of integers corresponding to the terms in the statement. *The Gödel numbers are not literally added together.* Rather the terms in the symbolic summation on the right side are individually rendered in a language[39] using the inverse relation of Gödel numbers $gi(g)$, (which is property 1 in section 3.1) on each term of the symbolic sum. Additional words such as "a" and "the" are added to the statement implicitly to make it grammatically correct.
        Thus we have a Hilbert space formalism for the semantics of a deductive system. We thus "turn the tables" and use the very well-known mathematics of Hilbert spaces as the "language" of our semantic

---

[39] Spaces and other required grammatical constructs such as "a" and "the" are assumed to be automatically inserted when translated into a language.

universe—rather than English or an artificial symbolic language, which both contain the possibilities of paradox and inconsistency.

One immediate benefit of a Hilbert space formulation is that each self-adjoint eigenvalue operator A_E has a corresponding projection operator[40] P_E that projects a composite state

$$|E_{tot}> = a|E> + b|0> \qquad (3.2.4a)$$

into a specific eigenvalue state $|E>$ or $|0>$: [41]

$$P_E|E_{tot}> = a|E> \qquad (3.2.4b)$$

In addition the operator $1 - P_E$ projects a state $|E>$ to 0. Thus

$$(1 - P_E)|E_{tot}> = b|0> \qquad (3.2.4c)$$

The projection operators for eigenvalue states are combined with semantic status projection operators P'_E that are used to determine whether a statement has a truth value:

$$P_{Etot} = P'_E P_E \qquad (3.2.5)$$

The set of these projection operators $\{P_{Etotk}, (1 - P_{Etotk})\}$, becomes the semantic status operators discussed in section 1.7. *All projection operators in a universe of discourse in this chapter commute.* The projection operator expression corresponding to eq. 3.2.3 (that determines whether the statement has a truth value) is

$$F(P'_{E_1}, P'_{E_2}, \ldots) \prod_{n \,\in\, \{m\}} P_{E_n} \qquad (3.2.6)$$

where $F(P'_{E_1}, P'_{E_2}, \ldots)$ will be seen to be the primary determinant of the truth value of the statement. The product of the projections P_{E_n} will be non-zero for the universe state (defined later in eq. 3.4.3), and zero if any

---

[40] Mackey (1963) pp. 64-65.
[41] The state $|0>$ corresponds to the empty string.

of the eigenvalues of P_{E_n} for $n \in \{m\}$ is zero.[42] The form of $F(P'_{E_1}, P'_{E_2}, ...)$ depends on the logical constructs ("and", "or", and so on) contained in the statement. We will see explicit examples of $F(P'_{E_1}, P'_{E_2}, ...)$ later.

Each projection operator filters a statement step by step

$$F(P'_{E_1}, P'_{E_2}, ...) \prod_{n \in \{m\}} P_{E_n} | E_1, E_2, E_3, ... , E_n > \qquad (3.2.6a)$$

to determine whether the statement has a truth value (a non-zero value) or not (a zero value). Eq 3.2.6 is the Operator Logic equivalent of measurement operator products such eq. 2.3.6 that filter the stages of an experiment. Measurement operators are also multiplicative projection operators.

3.2.2 The Calculus Universe of Discourse

A *calculus universe of discourse* consists of a set of primitive terms, and terms defined in terms of them, that are unrelated to any reality and have only a symbolic significance. They are used to formulate the axioms and theorems of a deductive system. A calculus may correspond to one or more semantic deductive systems. The formulation of a calculus universe of discourse is analogous to that of a semantic universe of discourse.

The terms (consisting of one or more symbols) in the calculus are mapped to Gödel numbers. Each term E has a unique Gödel number $gn(E)$. Each Gödel number is an eigenvalue of a unique, self-adjoint eigenvalue operator that we denote B_E which has two eigenvalues $gn(E)$ and 0. To each of these eigenvalues there corresponds eigenstates denoted $|E>$[43] and $|0>$ respectively. *Although we use the same notation as the previous subsection the calculus universe of discourse has different eigenvalues, different eigenstates and a different Hilbert space.* Eigenstates have the form:

$$|E_1, E_2, E_3, ... , E_q > = \prod_{m=1}^{q} |E_m > \qquad (3.2.7)$$

---

[42] Universe states containing some zero eigenvalues occur in the case of subuniverse states, which are described later.

[43] Again E can be a string or its corresponding Gödel number.

where q is the number of primitive terms and where each E_i can be a Gödel number or zero. The set of eigenstates defines a representation of the calculus Hilbert space. The terms, E_m, can have eigenvalues $gn(E_m)$ or $0 = gn("")$.[44] Note: The number of primitive terms, q, is finite since the number of symbols in a calculus is always finite by assumption. Each eigenvalue operator B_{E_k} satisfies the equation

$$B_{E_k}|E_1, E_2, E_3, \ldots ,E_q> = gn(E_k)|E_1, E_2, E_3, \ldots ,E_q> \qquad (3.2.8)$$

for $k \, \varepsilon \, [1, q]$. If $E_k = ""$ then $gn(E_k) = 0$. *All eigenvalue operators in a universe of discourse in this chapter commute.*

A calculus statement in this notation has the form

$$\sum_{n \, \in \{m\}} B_{E_n} \qquad (3.2.9)$$

where $\{m\}$ is a set of integers corresponding to the terms in the statement. If applied to an eigenstate we obtain the symbolic sum

$$\sum_{n \, \in \{m\}} B_{E_n} |E_1, E_2, E_3, \ldots ,E_q> = \sum_{n \, \in \{m\}} gn(E_n)|E_1, E_2, E_3, \ldots ,E_q> \qquad (3.2.9a)$$

It can be translated to a language[45] using the inverse relation of Gödel numbers $gi(g)$, (which is property 1 in section 3.1) on each term of the symbolic[46] sum in eq. 3.2.9 with appropriate grammatical constructs such as "a" and "the" added.

Thus we have a Hilbert space formalism for the calculus of a deductive system. We again use the mathematics of Hilbert spaces as the "language" of our calculus—differing from Gödel by using a formulation based on operators in a Hilbert space rather than Gödel numbers directly.

---

[44] The state $|0>$ corresponds to the empty string.

[45] Spaces and other required grammatical constructs such as "a" and "the" are assumed to be automatically inserted when translated into a language.

[46] Symbolic because addition is not performed. Rather each Gödel number is translated to its string equvalent.

Each self-adjoint eigenvalue operator B_E has a corresponding projection operator[47] Q_E that projects a state

$$|E_{tot}> = a|E> + b|0> \qquad (3.2.10)$$

into a specific eigenvalue state $|E>$ or $|0>$:

$$Q_E|E_{tot}> = a|E> \qquad (3.2.11a)$$

In addition the operator $1 - Q_E$ projects a state with eigenvalue gn(E) into $|0>$

$$(1 - Q_E)|E_{tot}> = b|0> \qquad (3.2.11b)$$

We define the projection operator

$$Q_{Etot} = Q'_E Q_E \qquad (3.2.12)$$

where Q'_E represents the equivalent of the semantic status projection operators and is used to determine whether a calculus statement is "well-formed." The set of these projection operators $\{Q_{Etotk}, (1 - Q_{Etotk})\}$ becomes the calculus status operators, which we will discuss shortly. *All projection operators in a calculus universe of discourse in this chapter commute.*

3.2.3 Variables

If a primitive term is in fact a function of one or more variables $x_1, x_2, \ldots$, then the corresponding eigenvalue operator and projection operator are also functions of these variables: $A_E(x_1, x_2, \ldots)$ and $P_{Etot}(x_1, x_2, \ldots)$ respectively for semantic universes of discourse; and $B_E(x_1, x_2, \ldots)$ and $Q_{Etot}(x_1, x_2, \ldots)$ for calculus universes of discourse. If a free variable v appears in a statement, then it has an eigenvalue operator $A_v(v)$ or $B_v(v)$, and a projection operator, $P_v(v)$ or $Q_v(v)$, for the respective universes of discourse. A function f(x) has an eigenvalue operator $A_f(x)$ or $B_f(x)$ in the respective universes with projection

---

[47] Mackey (1963) pp. 64-65.

operators $P_f(x)$ or $Q_f(x)$. The generalization to functions of several values is direct.

3.3 Linking Discourse Hilbert Spaces

The calculus and semantic Hilbert space statements are linked using a generalization of Tarski's method of combining a calculus sentence with its semantic meaning through eq. 1.5.1:

$$\text{SM is true if, and only if, S} \qquad (1.5.1)$$

as a starting point where S is a statement and SM is its semantic equivalent. If S is true, then SM is true. If S is false, then SM is false. If S is Gödel undecidable, then SM can be true or false.

Clearly, statement by statement, we can map semantic statements in a deductive system to syntactic statements in its corresponding calculus. This mapping consists of matching statements of the form of eq. 3.2.3 to statements of the form of eq. 3.2.9. It is clear that we can always define a *perfect map* on the operator level in which eigenvalues and their corresponding eigenvalue operators map on a one-to-one basis

$$A_{E_k} \leftrightarrow B_{E_k} \qquad (3.3.1)$$

Similarly, the projection operators also then map in a one-to-one fashion:

$$P_{E_{totk}} \leftrightarrow Q_{E_{totk}} \qquad (3.3.2)$$

Lastly, eq. 1.5.1 is true for the product of the semantic projection operators and the calculus projection operators of the corresponding statements. To be precise, consider a semantic statement with the form of eq. 3.2.3. Its corresponding product of projection operators is

$$P_S = F(P'_{E_1}, P'_{E_2}, \ldots) \prod_{n \in \{m\}} P_{E_n} \qquad (3.3.3)$$

The product of projection operators for the corresponding calculus statement is

$$Q_S = F(Q'_{E_1}, Q'_{E_2}, \ldots) \prod_{n \in \{m\}} Q_{E_n} \qquad (3.3.4)$$

The operator equivalents of the comments at the beginning of this section are:

1. Q_S is non-zero if, and only if, P_S is non-zero (indicating that the statement can be proven true or false, or provable or not provable, in both universes of discourse.)
2. Q_S is zero if, and only if, P_S is zero (indicating that the statement is Gödel undecidable in both universes of discourse – it cannot be proved and cannot be disproved.) However, the statement may be shown true or false (or provable or not provable) by other means – for example, from experiments.

where it is assumed that these projection operator expressions are evaluated between universe states defined later in eq. 3.4.3.

The reader will note that these points differ from the Tarski formulation (eq. 1.5.1) in several ways. Tarski's formulation assumes that the statements, S and SM, are either true or false – contrary to the implications of Gödel's Undecidability Theorem. Points 1 and 2 above are consistent with the Undecidability Theorem. Item 2 provides a calculational approach to determine "decidability."

English Language Statement	Semantic System Statement	Calculus Statement
Eigenvalue Operator Sum	Semantic Eigenvalue Operator Sum	Syntax Eigenvalue Operator Sum
Status Projection Operator Product	Semantic Projection Operator Product	Syntax Projection Operator Product

Figure 3.1 A table of the syntactical and semantic operators associated with a statement.

In chapter 1 (eqs 1.7.1 – 1.7.4) we considered a simple example of a two statement semantic system, which is also its calculus:

(1) This sentence is false.
(2) This sentence is true.

Statement (1) is undecidable – being neither true nor false. Statement (2) is true. The table (eq. 1.7.4) does not differentiate between the semantic system and the calculus (since they are essentially identical), and thus does not convey the full picture that we are developing.

The general case of the operator representation of an English (or other) language statement in a semantic system and its corresponding calculus statement is displayed in Fig. 3.1.

3.3.1 Predicate Calculus

Returning to an earlier example we see that the predicate calculus can be implemented in our Hilbert space representations. First we note the statement generated from the Subject and Predicate operators is

$$((\text{Subject}) + (\text{Predicate}))|\Omega_1> = \text{"The car is red"} \,|\Omega_1> \qquad (3.3.5)$$

where again we note it is implemented via corresponding self-adjoint operators with Gödel number eigenvalues.

Correspondingly one can define projection operators (as in chapter 1):

$$P_{\text{Subject("The car")}} = P_1$$

and $\qquad\qquad\qquad\qquad\qquad\qquad\qquad\qquad\qquad\qquad (3.3.6)$

$$P_{\text{Predicate("is red")}} = P_2$$

(The simplicity of the example allows us to treat semantic and calculus operators similarly.) The product of the projection operators for the statement, when evaluated for the universe state (eq. 3.4.3 below) is not identically zero, and so the statement has a truth value of true or false.

This simple example can be directly extended to more complex statements with multiple subjects and clauses.

3.3.2 General Approach

In a given universe of discourse we can introduce a projection operators P_i for each predicate, or for primitive terms within a predicate as the case may be. Then, for each predicate (predicate i – denoted P_i), we determine the class of acceptable subjects (subjects, which when combined with the predicate to form a statement, form a valid statement that is either true or false). We can assign a projection operator P_i to each member of the class of allowed subjects. Unacceptable subjects (subjects, which, when combined with P_i to form a statement, form an invalid statement that has no truth value) are assigned the projection operator $1 - P_i$. Thus, in a universe of subjects and predicates, each predicate has a class of acceptable subjects (its *domain* of allowed subjects) and a complementary class of unacceptable subjects.[48]

Statements formed from a given predicate with an unacceptable subject are either nonsense statements or paradoxical statements of the type discussed earlier and will be analyzed later.

3.4 Measurement Operator Formalism for Operator Logic

In this section we will look at the operator formalism for a semantic universe of discourse. A *statement* is represented by an inner product[49] containing a sum of the eigenvalue operators of the statement:

$$\text{statement} = <\Omega_u|A_1 + A_2 + A_3 + \dots |\Omega_u> \qquad (3.4.1)$$

and its truth value[50] is given by

$$TV = \text{Truth Value} = <\Omega_u| \, F(P'_{E_1}, P'_{E_2}, \dots) \prod P_{E_n} |\Omega_u> \qquad (3.4.2)$$

where $TV = 0$ indicates the statement has no truth value, where $TV \neq 0$ indicates the statement has a truth value: true or false, and where A_i is the ith eigenvalue operator. The projections appearing in F and the product are those of the eigenvalue operatos in eq. 3.4.1.

We will define the universe state $|\Omega_u>$ for a universe with q primitive terms as

$$|\Omega_u> = |E_1, E_2, E_3, \dots, E_q> \qquad (3.4.3)$$

---

[48] In the case of several predicates in a universe of discourse each predicate will have its own domain of subjects. Then each truth value projection operator becomes a product of commuting projection operators – one factor for each predicate – so that any combination of subject and predicate yields a truth value or has no truth value.
[49] Expressions of the form $<\alpha|O|\beta>$ are inner products that are called *expectation values* in quantum mechanics and are often denoted as $(\alpha, O\beta)$ in mathematics texts where O is an operator.
[50] In the case of a calculus universe of discourse truth value becomes provability.

where each eigenvalue E_i is a non-zero Gödel number and the set of integers 1, 2, ..., q is denoted {m}. We call this state the **universe state** of the universe. Its hermitean conjugate is denoted $<\Omega_u|$. Eqs. 3.4.1 – 3.4.3 (and the following equations) use Dirac's bra-ket notation for states and inner products.

A statement is composed of adjoining (a symbolic sum of) Gödel numbers: gn(i). Using the inverse function gi(gn(i)) on each adjoining Gödel number (separated by symbolic plus signs) the statement can be "translated" into English or some other human or symbolic language. The plus signs act as separators and are subsequently discarded. Grammatical constructs such as "a" and "the" can be added as needed to make a grammatically correct statement.

3.5 An Example

One example of such a semantic universe has the primitive terms (and corresponding eigenvalue operators):

Primitive Term	Eigenvalue Operator	Status Projection Operator
"This sentence"	A_1	P_1
"Some Sentence"	A_2	P_2
"is true"	A_3	$P_3 = P_1$
"is false"	A_4	$P_4 = 1 - P_1$

where we assign projection operators to each term using an operator P defined below. The primitive terms consist of two predicates, "is true" and "is false". The other primitive term is the subject: "This sentence". The operator $P_2 = P$ projects an eigenstate as follows

$$P_1|\Omega_1> = |"This sentence", "Some sentence", "is true", "is false">$$

while

$$(1 - P_1)|\Omega_1> = 0$$

and

$$(1 - P_1)| "", "Some sentence", "is true", "is false"> =$$
$$= | "", "Some sentence", "is true", "is false">$$

We can construct four simple statements from these primitive terms:

English Statement	Eigenvalue operator Sum	Projection Operator Inner Product $<\Omega_u|\text{product}|\Omega_u>$
"This sentence is true"	$A_1 + A_3$	$P_1P_3 = P_1 \neq 0$
"This sentence is false"	$A_1 + A_4$	$P_1P_4 = P_1(1 - P_1) \equiv 0$
"Some sentence is true"	$A_2 + A_3$	$P_2P_3 = P_2P_1 \neq 0$
"Some sentence is false"	$A_2 + A_4$	$P_2P_4 = P_2(1 - P_1) \neq 0$

where the zero and non-zero values in the third column result from taking the inner product $<\Omega_1|\text{product}|\Omega_1>$.

From the preceding table we see that the statements represented by $A_1 + A_3$, $A_2 + A_3$ and $A_2 + A_4$ have truth values (either true or false). The statement represented by $A_1 + A_4$ does not have a truth value.

The statements represented by $A_1 + A_3$, $A_2 + A_3$ and $A_2 + A_4$ have the truth value true from semantic considerations. This result is not due to the projection operator results. Projection operators only determine whether a statement has a truth value, or not, *at this point in our discussion.*[51]

For each of the predicates A_3 and A_4 we identified a domain of subjects that lead to statements that have a truth value. For A_3 we find A_1 and A_2 are subjects that leads to statements with a truth value and thus constitute the domain of A_3. In the universe specified above we find A_4 has a domain of A_2.

It is clear that the identification of the domain of subject terms for each predicate that leads to statements with a truth value is the key to excluding paradoxical statements. We will see this in more detail after discussing primitive terms in general in our formulation. This discussion can be directly generalized to more complex statements.

3.6 The Primitive Terms of Universes of Discourse

The set of primitive terms of a universe of discourse are the "string eigenvalues" that are mapped to the Gödel numbers of the eigenvalue operators of the universe. Each primitive (and defined) term

---

[51] At a later point in the development they will be used to determine the truth or falsity of statements.

maps to one eigenvalue operator. Then all statements in the universe of discourse can be expressed as symbolic sums of the corresponding eigenvalue operators.

The primitive terms are the key to the entire universe of discourse. We now consider primitive terms in human and symbolic languages.

Statements can be expressed in primitive terms in human languages such as English or in artificial symbolic languages. A problem that arises—particularly in human languages—is that there are different ways of expressing the same statement and different ways of specifying the same primitive terms. A particularly good example of this issue appears when one compares a primitive term in German, which can create words by combining other words, and English where the equivalent of a German primitive term may be a phrase.

In this section we point out that one can define a map between the primitive terms of two semantic universes of discourse that have the same calculus universe of discourse. Thus no problem can arise if we choose to use differing primitive terms in two semantic universes of discourse with the same calculus universe of discourse. A famous example of this situation is the calculus of Euclidean geometry *without* the fifth postulate. Depending on the definition of "point", "straight line" and "congruence" one obtains Euclidean geometry or a non-Eucliean geometry.[52]

Thus the issue of primitive terms is important in the consideration of semantic universes of discourse. We will now consider general features of primitive terms.

A universe of discourse consists of

1. Primitive terms
2. Axioms expressed in terms of the primitive terms
3. Terms defined as combinations of primitive terms
4. A set of theorems derivable from the axioms

The primitive terms in a calculus universe of discourse are undefined. The primitive terms in a semantic universe of discourse are given a

---

[52] Weyl (1950) p. 81.

meaning external to the universe of discourse. Frequently the meaning is based on physical or mathematical entities.

If two semantic universes of discourse have the same calculus universe of discourse then the primitive terms in every semantic universe must map to each other on a one-to-one basis. (It is possible that a primitive term in one universe of discourse maps to a pair of (or several) primitive terms in the other universe of discourse. However the pair of (or several) terms always appear together in the other universe of discourse with the result that they are effectively one term.)

3.7 Predicate Domains and Paradox Avoidance

From the discussion of previous sections it is clear that each predicate has a domain of allowed subjects.[53] The subjects that are not in the domain of allowed subjects can be used to form statements. But these statements would have no truth value.[54] They would be neither true nor false and thus would properly be called undecidable or paradoxical statements.[55]

We described most of the well-known paradoxes in chapter 1. We will now reconsider them in the light of our new formalism.

Liar Paradox

The statement of the Liar Paradox has a subject such as Subject = "The statement that I am now making" and predicate "is a lie." If the subject is reflexive as it appears implicitly in the predicate, which might be better phrased as "is my lie", then a paradox is evident. The elimination of this paradox is accomplished by limiting the domain of the predicate to non-reflexive subjects. In the Operator Logic formalism the domain of the predicate operator is limited to non-reflexive subject operators. One implements this by assigning a projection P to the

---

[53] In this respect our view is contrary to Leibniz's view of 1686, and other views, that the predicate is contained in the concept of the subject in each true statement. Our view does agree with Frege's use of a functional notation wherein the predicate is the function and a subject is the variable. The set of subjects constitutes the domain of the function. Further, it agrees with Hilbert (1950), "To each predicate corresponds a certain 'class' of objects, consisting of all objects for which the predicate holds." p. 46.

[54] Truth value becomes provability in a calculus universe of discourse.

[55] Herbrand, Schmidt, Wang, Hintikka, Hailperin, Lightstone & Robinson, Gilmore, and Quine have considered approaches within the framework of earlier formalisms based on restricting the subject domain.

predicate and the projection $(1 - P)$ to all reflexive subject operators. Thus paradoxical statements of this type have no truth value and the Operator Logic formalism excludes such paradoxes.

Grelling Paradox

Some adjectives have the property that they apply to themselves. Such adjectives can be called autological. Other adjectives do not apply to themselves. These adjectives can be called heterological. The Grelling paradox is:

"Heterological is heterological."

If we identify the primitive terms as the predicate "is heterological" and the subject "Heterological", then assigning eigenvalue operators to each term, and the projection operator P as the subject projection operator, and $(1 - P)$ as the predicate projection operator gives the Grelling statement no truth value, thus eliminating the paradox.

Barber Pseudoparadox

The Town mayor issues an order: "The one town barber must shave those men in the village who do not shave themselves." In the simplest Operator Logic form the subject is "The one town barber" and the predicate is "must shave those men in the village who do not shave themselves." Assigning eigenvalue operators to each term and the subject projection operator P with the predicate projection operator $(1 - P)$ results in the statement not having a truth value. Thus the paradox is avoided.

Berry Paradox

The number of positive integers that can be named in English in less than a fixed number of syllables is finite. Thus there must be a least integer that cannot be so named. However, " the least integer that cannot be named in English in less than fifty syllables" is an English name of less than fifty syllables. Thus the least integer has a name contrary to the assumption and thus a paradox.
Resolution:

This superficial paradox is based on an ambiguity in the use of the word "named." If it means "numerically named" (in words representing a numerical value) there is no paradox. If it means "any type of name" then an apparent paradox appears until one realizes that it is then a tautology "The least integer that cannot be named in English in less than fifty syllables is the least integer that cannot be named in English in less than fifty syllables."

Russell's Paradox

From experience we know we can consider classes of things; classes of integer numbers, classes of cars, and so on. We can consider classes of classes such as the class of all classes of cars in the various big cities of America. There are two interesting varieties of classes: proper classes and improper classes. *Proper classes* are classes, which are not members of themselves. For example, the class of all cars in China is proper. *Improper classes* are classes, which are members of themselves. For example, the class of all classes is a member of itself and thus improper.

Let us define the Russell class R as the class of all proper classes. Paradox: if R is a proper class, it is a member of itself, and is thus by definition an improper class. If R is an improper class, it is not a member of itself and therefore, by definition, it is a proper class. Thus there is no resolution of this paradox according to conventional logic.

Resolution:

The primitive terms of this paradox are the subject, "the class of all proper classes" and the predicate, "is a proper class." After associating an eigenvalue operator and projection operator with each term we see that the statement "the class of all proper classes is a proper class" has no truth value if we specify a projection operator P for the subject and $(1 - P)$ for the predicate. This is again a case where the paradox is avoided by the predicate having a non-reflexive domain.

Cantor Paradox

According to the theory of cardinal (infinite) numbers the set of all subsets of a set C has a higher cardinal number than C. If C is the set of all sets, then the preceding statement is a contradiction.

Resolution:

The Cantor paradox can be changed to the following by simple substitution for C:

The set of all subsets of the set of all sets has a higher cardinal number than the set of all sets.

The subject can be simplified to give:

The set of all sets has a higher cardinal number than the set of all sets.

which is a manifestly verbal paradox but not an Operator Logic paradox.

If we identify the subject term as "The set of all sets" and the predicate term as "has a higher cardinal number than the set of all sets", and define corresponding eigenvalue and projection operators, then the statement can be shown to have no truth value eliminating the paradox. We choose the projection operator for the subject as P and the projection operator for the predicate as $(1 - P)$ with the result that the statement has no truth value within the framework of Operator Logic. Again the paradox is avoided by the predicate having a non-reflexive domain.

Burali-Forti Paradox

The Burali-Forti paradox is based on the theory of transfinite (infinite) numbers. The theory proves a) every well-ordered set has a unique ordinal number; b) any set of ordinals, that is placed in a natural order such that each element contains all its predecessors, has an ordinal number which is greater than any preceding element in the set; and c) the set A of all ordinals placed in natural order is well-ordered. Then by theorems a and c, A has an ordinal number n. Since n is in A we see n < n by theorem b, thus establishing a contradiction.

Resolution

The paradoxical conclusion can be stated verbally as:

The ordinal of the set of all ordinals is greater than the ordinal of the set of all ordinals.

In this statement we can identify the subject primitive term as "The ordinal of the set of all ordinals" and the predicate primitive term as "is greater than the ordinal of the set of all ordinals." Here again we see a reflexivity in the subject. If we restate the conclusion in terms of eigenvalue operators with the subject projection operator P and the predicate projection operator $(1 - P)$ then the inner product $<\Omega_u|P(1 -$

P)$|\Omega_u>$ = 0 indicating that the statement has no truth value. Thus the statement is excluded from the set of statements with truth values in this universe of discourse calculus. The domain of the predicate does not include reflexive subjects such as the one given above.

Richard Paradox

The Richard paradox is concerned with the proposition: the set of all numerical functions is not enumerable. A commonly used argument to prove this proposition is the following. Suppose an enumeration existed symbolized by $f_n(m)$, which represents the nth function with argument m. Consider the function g defined by

$$g(n) = f_n(n) + 1$$

for any value of n. Let n_0 be the index number of g(n) in the enumeration:

$$g(n) = f_{n_0}(n) + 1$$

then

$$f_{n_0}(n) = g(n) = f_n(n) + 1 \qquad (3.7.1)$$

and

$$f_{n_0}(n_0) = g(n_0) = f_{n_0}(n_0) + 1 \qquad (3.7.2)$$

which is a contradiction. Thus the set of all numerical functions is not enumerable.

Contrarian argument: Consider the set of all definable functions. Definable is taken to mean definable in some specific language with a fixed dictionary and grammar. Since the number of words in the language is finite, then the number of expressions is enumerable. Thus the set of expressions that form the definitions of definable functions is enumerable. Thus the set of definable functions is enumerable. Since the set of numerical functions is a subset of the set of definable functions it also must be enumerable. Thus the set of all numerical functions is enumerable.

The result of the preceding two arguments is a contradiction (paradox).

Resolution

The projection operator inner product corresponding to eq. 3.7.1 has the form (see subsection 3.2.3 for a discussion of variables)

$$<\Omega_u| \, \dots \, P_{f_{n_0}}(n) \, \dots \, (1 - P_{f_n}(n)) \, \dots \, |\Omega_u> \qquad (3.7.3)$$

which is non-zero (and thus the statement has a truth value) for $n \neq n_0$. However for $n = n_0$ eq. 3.7.3 is equal to zero[56] and thus eq. 3.7.2 has no truth value and therefore the paradox is avoided. The contrarian proof is not contradicted.

Gödel's Undecidability Theorem

Gödel's Undecidability Theorem, as he pointed out in his celebrated paper, is closely related to Richard's paradox (antinomy) and the Liar's paradox. Gödel provides an explicit undecidable statement, eq. 1, in his paper:

$$n \, \varepsilon \, K \equiv \overline{Bew}[R(n); n] \qquad (3.7.4)$$

where

1. Bew Y means Y is a provable formula (statement)
2. The bar over Bew means "not"
3. K is a class of natural numbers
4. A class-sign is a formula[57] with one free natural number variable
5. The set of class-signs is arranged in a series using some rule with the nth class-sign denoted R(n).
6. Define [α, m] to be the formula derived upon replacing the free variable in the class-sign α by the sign of the natural number m. Thus the relation Y = [α, z] is definable in PM.
7. Since the above are all definable in PM, so is the class K that is formed from them.

Therefore there is a class-sign S such that [S; n] implies that $n \, \varepsilon \, K$. Thus S = R(q) for some specfic natural number q. If we assume [R(q); q] is provable, then $q \, \varepsilon \, K$. But, by eq. 3.7.4

---

[56] Note all projection operators commute as do all eigenvalue operators in the formulation of this chapter.
[57] Of the Principia Mathematica (denoted PM).

$$\overline{Bew}[R(q); q] \qquad (3.7.5)$$

would be true contrary to our assumption.

If we assume the contrary: the negation of

$$[R(q); q] \qquad (3.7.6)$$

is provable, then

$$\overline{q \,\varepsilon\, K} \qquad (3.7.7)$$

and

$$Bew[R(q); q] \qquad (3.7.8)$$

would be true. Thus [R(q); q] and its negation would be provable.

Gödel's Theorem (more precisely any statement conforming to eq. 3.7.4) is excluded from a mathematical deductive system if the subject is reflexive. Eq. 3.7.4 can be stated in a subject predicate form as

"n is contained in K which is equivalent to …"

where "n" is the subject and "is contained …" is the predicate. In this form we can associate the projection P with "n" and $(1 - P)$ with the predicate part "[R(n); n]" yielding

$$<\Omega_u|P \ldots (1 - P)|\Omega_u> = 0 \qquad (3.7.9)$$

which shows the statement has no truth value.

Gödel's more succinct undecidable statement[58] 17 Gen r is similar in character but disguised in a more condensed notation. The observant reader will note the close resemblance between Gödel's undecidablity result, Richard's paradox and the Liar paradox—as Gödel himself noted.[59]

Conclusion

The preceding examples show that Operator Logic provides a mechanism to systematically exclude statements with no truth value

---

[58] Gödel (1992) p. 60.
[59] Gödel (1992) p. 40.

("undecidable" or paradoxical statements) from a universe of discourse. The following chapters show further advantages of Operator Logic including quantum probabilistic logic, hierarchies of universes of discourse, and the formulation of a mathematical philosophy (metaphysics) relating Platonic Ideas to the physical universe, Reality, that is conceptually similar to theories of various Platonic schools.

3.8 Sentential Calculus in Operator Logic

In this section we will extend our discussion of projection operators to determining the truth value of simple sentences, or clauses, consisting of a subject and a predicate.

In their classic book, Hilbert and Ackermann[60] begin with a discussion of the sentential calculus[61] (calculus of sentences[62]). They point out that sentences can be combined in a number of ways to create compound sentences using connectives such as "and", "or", "if ... then", and "not". They specify five fundamental combinations of sentences using the following symbols:

Connective	Symbol
and	&
or	v
if ... then (modus ponens)	$\rightarrow$
if and only if	~
not[63]	___

The symbols can be used to form fundamental combinations of sentences.

Denoting sentences with upper case letters: A, B, ... the five basic combinations are:

1. A & B is a compound sentence that is true if and only if both A and B are true.
2. A v B is true if and only if at least one of the sentences A or B is true.

---

[60] Hilbert (1950).
[61] We assume a semantic universe of discourse in this, and the following, subsections.
[62] We use sentence and statement interchangeably.
[63] For lexicographic reasons we use underscore ___ for "not" rather than the more commonly used overscore ‾ symbol.

3. $A \rightarrow B$ is false if and only if A is true and B is false.
4. $A \sim B$ is true if and only if both A and B are true, or both A and B are false.
5. The sentence $\underline{A}$ is false if A is true, and is true if A is false.

3.8.1 Truth Value of Simple Sentences and Clauses

The projection operators, that we have used up to now, have determined the decidability of sentences or clauses. A *clause* consists of a subject(s) and predicate. A sentence or clause was undecidable if the product of the subject projection operator and predicate projection operator was zero—usually in the simple form of $P(1 - P) = 0$.

Now we extend the formalism to determining the truth value of a sentence or clause using these projection operators. The procedure begins by assigning a set of projections, which we will call *truth projections*, to each predicate—one truth projection for each subject in the domain of subjects for the predicate. For predicate j and subject i we will denote the truth projection as P_{pij}. For each subject i we associate a projection P_{si}. Then the sentence

$$<\Omega_u|A_1 + A_2|\Omega_u> \tag{3.8.1}$$

where A_1 is the subject and A_2 is the predicate, has the truth value

$$\text{Truth Value} = <\Omega_u|P_{s1tot}P_{p12tot}|\Omega_u> \tag{3.8.2}$$

For example, if the sentence is "The sky is blue" then when we assign truth projections for all possible subjects of the predicate "is blue" we could assign

$$P_{s1tot} = P_{p12}$$

and normalize P_{p12} such that $<\Omega_u|P_{p12}|\Omega_u> = 1$ with the result that the truth value is +1 which we interpret as true.

We assign numeric values to truth values in this formalism for Classical Operator Logic:

$$
\begin{aligned}
\text{True} &= +1 \\
\text{Undecidable} &= 0 \\
\text{False} &= -1
\end{aligned}
\tag{3.8.3}
$$

If, on the other hand, the sentence is "The sky is green" then when we assign truth projections for all possible subjects of the predicate "is green" we would assign

$$P_{s1tot} = P'_{s1}P_{p12} = -P_{p12}$$

so that the truth value would be -1 meaning false.

The observant reader will note that the above choice of truth projection operator satisfies $P_{s1tot}P_{s1tot} = -P_{s1tot}$ and is thus not a projection of the normal sort. Therefore, in order to accommodate negative values, which indicate false, we will generalize truth projection operators to satisfy

$$P^2 = \pm P \qquad (3.8.3a)$$

For a given predicate, the choice of subject truth projections is a semantic issue and thus is determined by circumstances.

If a given sentence has multiple subjects the determination of the truth value is slightly more complex. We consider the cases:

$$\text{Subject1 and subject2 predicate} \qquad (3.8.4)$$

and

$$\text{Subject1 or subject2 predicate} \qquad (3.8.5)$$

In the case of eq. 3.8.4 the truth value[64] is obtained by evaluating the expression:[65]

$$\text{Truth Value} = <\Omega_u|P_{s1}P_{p12}|\Omega_u><\Omega_u|P_{s2}P_{p12}|\Omega_u>[1 - \\ - 2\theta(- <\Omega_u|P_{s1}P_{p12}|\Omega_u> - <\Omega_u|P_{s2}P_{p12}|\Omega_u>)] \qquad (3.8.6)$$

---

[64] Expressions of the form $<\alpha|O|\beta>$ are inner products that are called *expectation values* in quantum mechanics and are often denoted as $(\alpha, O\beta)$ in mathematics texts. In this example and the following examples we factor expectation values into products of expectation values of clauses since this procedure is both well defined and produces results consistent with our intuitive expectations. Interestingly, a somewhat similar procedure is followed in perturbation theory expansions in quantum field theory. In perturbation theory, terms with more than two operators in them are expanded in terms of vacuum expectation values of products of two quantum field operators.

[65] The statement is separated into an equivalent consisting of two parts each of which is evaluated using the universe state $|\Omega_u>$. It is clearly the only way of obtaining truth values correctly in this formalism.

where
$$\theta(x) = 1 \text{ if } x > 0 \text{ and } \theta(x) = 0 \text{ if } x \leq 0 \qquad (3.8.7)$$

since the subject-predicate expression must be true for both subjects for the statement to be true. If either factor in eq. 3.8.6 is zero then the truth value of the statement is zero meaning it is undecidable. If either factor, or both factors, in eq. 3.8.6 are false (and both decidable) then the truth value of the statement is –1 meaning false.

In the case of eq. 3.8.5 the truth value is obtained by evaluating the truth projection expression:

$$\text{Truth Value} = \text{Max}(<\Omega_u|P_{s1}P_{p12}|\Omega_u>, <\Omega_u|P_{s2}P_{p12}|\Omega_u>) \cdot$$
$$\cdot |<\Omega_u|P_{s1}P_{p12}|\Omega_u><\Omega_u|P_{s2}P_{p12}|\Omega_u>| \qquad (3.8.8)$$

since the statement is true if the subject-predicate expression is true for either subject. The absolute value $|<\Omega_u|P_{s1}P_{p12}|\Omega_u><\Omega_u|P_{s2}P_{p12}|\Omega_u>|$ guarantees the truth value of the statement is zero if either or both subject-predicate expressions is undecidable. The function Max(a, b) is the maximum of the quantities a and b.

The generalization to the case of multiple subjects is direct.[66]

3.8.2 Truth Value of Compound Sentences and Clauses

In the case of compound sentences and clauses, classical Operator Logic is based on factoring them into simple subject-predicate clauses as discussed in the previous subsection.

In this subsection we will consider some general cases of compound sentences to illustrate the procedure for obtaining a truth value.

Suppose we have a sentence that can be expressed as a series of "and" clauses such as

---

[66] In the case of quantifiers such as "all" or "some" we make the quantifier part of the subject or predicate as the case may be. For example, in the sentence "All men are good" the subject is "all men" and the predicate is "are good". The truth value status of the sentence is determined by whether "all men" is a member of the domain of "are good". A similar example could be created for "some". And the appearance of "all" or "some" in a predicate also can be similarly treated. For example, "Peacemakers are all good men" has the predicate "are all good men" and the truth value status of the sentence is determined by whether "peacemakers" is in the domain of that predicate. Other quantifiers such as "there exists" and "for any" are embodiable within Operator Logic.

$$\text{Statement} = \text{clause1 and clause2 and ...} \tag{3.8.9}$$

Then the truth value of the statement is given by the expression:[67]

$$<\Omega_u|\text{statement}|\Omega_u> = |<\Omega_u|\text{clause1}|\Omega_u><\Omega_u|\text{clause2}|\Omega_u>...|\{1 -$$
$$- 2\theta(|<\Omega_u|\text{clause1}|\Omega_u>| - <\Omega_u|\text{clause1}|\Omega_u> + |<\Omega_u|\text{clause2}|\Omega_u>| - <\Omega_u|\text{clause2}|\Omega_u> +$$
$$+ ...)\} \tag{3.8.10}$$

Note the following truth value cases implied by eq. 3.8.10:

Case 1: all clauses true (no undecidable clauses) – Truth Value = +1
Case 2: One or more false clauses (with no undecidable clauses) – Truth Value = −1
Case 3: One or more undecidable clauses – Truth Value = 0

Suppose we consider a sentence that can be expressed as a series of "or" clauses such as

$$\text{Statement} = \text{clause1 or clause2 or ...} \tag{3.8.11}$$

Then the truth value of the statement is given by the expression:

$$<\Omega_u|\text{statement}|\Omega_u> = [2\theta(\theta(<\Omega_u|\text{clause1}|\Omega_u>) + \theta(<\Omega_u|\text{clause2}|\Omega_u>) + ...) - 1]\cdot$$
$$\cdot|<\Omega_u|\text{clause}|\Omega_u><\Omega_u|\text{clause}|\Omega_u><\Omega_u|\text{clause}|\Omega_u>...| \tag{3.8.12}$$

The following truth value cases are implied by eq. 3.8.12:

Case 1: Any clauses are true (no undecidable clauses) – Truth Value = +1
Case 2: All clauses are false (with no undecidable clauses) – Truth Value = −1
Case 3: One or more undecidable clauses – Truth Value = 0

Lastly, let us consider the case of an "if ... then" statement:

$$\text{If A, then B} \tag{3.8.13}$$

This statement is equivalent to

$$A\underline{B} \tag{3.8.14}$$

---

[67] Note that because the projection expressions do not change the universe state the expression $<\Omega_u|\text{clause1}|\Omega_u><\Omega_u|\text{clause2}|\Omega_u> \equiv <\Omega_u|\text{clause1clause2}|\Omega_u>$ and similarly elsewhere. The separation into "clause" factors is clearer.

which in words is NOT(A and NOT B). In terms of Operator Logic eqs. 3.8.13 and 3.8.14 can be represented numerically by

$$- [<\Omega_u|A|\Omega_u>(- <\Omega_u|B|\Omega_u>)] = <\Omega_u|A|\Omega_u><\Omega_u|B|\Omega_u> \qquad (3.8.15)$$

where $<\Omega_u|A|\Omega_u>$ and $<\Omega_u|B|\Omega_u>$ are evaluated in a manner similar to the previous cases eqs. 3.8.10 and 3.8.12 taking account of the particular form of A and B.

3.8.3 Truth Value of the Normal Form

It is well known that all sentential calculus statements can be reduced to operations involving two logical connectives such as & and __ (not).[68] So our Operator Logic clearly suffices to handle all sentential calculus statements including undecidable statements.

Furthermore any set of sentences can be into a form called a *normal form*[69] by means of equivalence transformations.[70]

Since our Operator Logic also encompasses the predicate calculus and second level predicate calculus (section 4.3 in the following chapter) we conclude we have a complete formalism[71] for classical Logic that decisively handles the issue of undecidability.

---

[68] Hilbert(1950) pp. 10-11.
[69] "consisting of a conjunction of disjunctions in which each component of the disjunction is either an elementary sentence or the negation of one" in the words of Hilbert (1950) p. 12.
[70] Hilbert(1950) p. 12.
[71] The rules of inference and proof procedures are analogous in Operator Logic and conventional Logic, and so will not be considered here.

4. Why Operator Logic?

4.1 Operator Logic is the Correct Formalism for Logic

Operator Logic enables logical systems and statements to be constructed that exclude paradoxes and undecidable statements as well as providing other important advantages described later. Thus it provides a superior formalism for Logic without the known defects of conventional language/symbolic language formulations, which we now see as incomplete.

4.2 Other Advantages of Operator Logic

Operator Logic has major advantages, which we will discuss in this chapter and in succeeding chapters:

1. Operator Logic provides a unified formulation of classical and quantum probabilistic logic.

2. Operator Logic is the only approach with a fundamental, non-intuitive basis in nature and thus is guaranteed to be consistent.[72]

3. Raising and lowering operators can be defined that enable us to create subuniverses of Calculus and of Semantic Universes of Discourse.

4. Projection operators can also be used to generate subuniverses.

---

[72] G. Feinberg, Phys. Rev. **159**, 1089 (1967) quotes Dr. M. Tausner to the effect, "No [experimental] observations can be logically inconsistent." Thus no formalism truly embodying Nature can be inconsistent. This comment presumes nature to be consistent and to embody a well-defined set of laws. The regularity of nature has become more and more evident in recent centuries with the growth of scientific knowledge.

5. A semantic universe of discourse can be defined, which can be mapped to a physical universe realizing the Platonic concepts of Ideas (pure forms), intermediate mathematical connections and the physical universe of objects, Reality.

These possibilities will be discussed in the following sections and chapters.

4.3 Logic Raising and Lowering Operators

4.3.1 Eigenvalue Operators that are not Functions of Variables

If we consider any eigenvalue operator (without a dependence on variables) A_E, and its corresponding eigenstates $|0>$ and $|E>$, then it is easy[73] to define raising and lowering operators that transform one eigenstate into the other. The raising operator is $a_E^\dagger$ and satisfies

$$|E> = a_E^\dagger|0> \qquad (4.3.1)$$

and, the lowering operator is a_E and satisfies

$$|0> = a_E|E> \qquad (4.3.2)$$

where † signifies hermitean conjugatation. In the present situation we do not wish repeated application of the raising operator to generate states with higher eigenvalues.[74] Therefore

$$a_E^\dagger a_E^\dagger|0> = 0 \qquad (4.3.3)$$
$$a_E a_E|E> = 0 \qquad (4.3.4)$$

Consequently the reader can verify that the raising and lowering operators must satisfy the anticommutation relations:

$$\{a_E^\dagger, a_E\} = gn(E)$$

---

[73] This is commonly done in the case of a harmonic oscillator in Quantum Mechanics, and also in Quantum Field Theory.
[74] Since they don't necessarily relate to primitive or defined terms.

$$\{a_E, a_E\} = 0 \qquad (4.3.5)$$
$$\{a_E{}^\dagger, a_E{}^\dagger\} = 0$$

for eqs. 4.3.1 – 4.3.4 to hold.[75]

More generally if there are a number of semantic eigenvalue operators A_{E_k} where $k = 1, 2, \ldots, q$ then the anticommutation relations are

$$\{a_{E_k}{}^\dagger, a_{E_{k'}}\} = gn(E_k)\delta_{kk'}$$
$$\{a_{E_k}, a_{E_{k'}}\} = 0 \qquad (4.3.6)$$
$$\{a_{E_k}{}^\dagger, a_{E_{k'}}{}^\dagger\} = 0$$

where $\delta_{kk'}$ is the Kronecker delta. The semantic eigenvalue operators can be represented in terms of these operators by

$$A_{E_k} = a_{E_k}{}^\dagger a_{E_k} \qquad (4.3.7a)$$

and the corresponding projection operators by

$$P_{E_k} = P'_{E_k} \tfrac{1}{2}[1 + [gn(E_k)]^{-1} a_{E_k}{}^\dagger a_{E_k}] \qquad (4.3.7b)$$

where P'_{E_k} is a truth projection operator. (See subsection 3.8.1 for its definition and use. The ' character in the P' string is omitted in section 3.8.1.) The product of P'_{E_k} operators determines the truth value of expressions including embodying the non-reflexivity of subjects and other criteria that preclude logical paradoxes.

The projection factor $\tfrac{1}{2}[1 + [gn(E_k)]^{-1} a_{E_k}{}^\dagger a_{E_k}]$ in eq. 4.3.7b projects states from the universe state whose k^{th} eigenvalue is zero. Thus the projection operator P_{E_k} plays two roles: 1) it projects a state $|\Omega>$, whose k^{th} eigenvalue is zero, to zero; and 2) it enables statements to avoid paradoxes by zeroing statements and clauses whose subjects are not in the subject domains of each predicate. (Subjects that are not in the subject domain of a predicate have a projection operator $P'_{E_{k'}}$ that when multiplied by the the predicate's P'_{E_k} gives zero identically. As a result

---

[75] This definition of raising and lowering operators leads to a simpler form for eq. 4.3.7a.

such a statement's corresponding projection operator product contains both projection operators thus yielding a zero truth value.)

Calculus eigenvalue operators, and raising and lowering operators have a similar form

$$\{b_{E_k}{}^\dagger, b_{E_{k'}}\} = gn(E_k)\delta_{kk'}$$
$$\{b_{E_k}, b_{E_{k'}}\} = 0 \qquad\qquad (4.3.8)$$
$$\{b_{E_k}{}^\dagger, b_{E_{k'}}{}^\dagger\} = 0$$

with

$$B_{E_k} = b_{E_k}{}^\dagger b_{E_k} \qquad\qquad (4.3.9a)$$

and

$$Q_{E_k} = Q'_{E_k}{}^{\frac{1}{2}}[1 + [gn(E_k)]^{-1} b_{E_k}{}^\dagger b_{E_k}] \qquad (4.3.9b)$$

where Q'_{E_k} is the equivalent of a semantic truth projection operator. The Q'_{E_k} projection operator prevents paradoxes in a manner similar to the semantic case. It embodies the non-reflexivity of subjects and other criteria that preclude logical paradoxes.

4.3.2 Eigenvalue Operators that are Functions of Variables

If we consider any eigenvalue operator that is dependent on free variables such as $A_E(x_1, x_2, ...)$ with corresponding eigenstates $|0, x_1, x_2, ...>$ and $|gn(E), x_1, x_2, ...>$ for each set of values of $x_1, x_2, ...$ then we can again define operators that transform between eigenstates. The raising operator becomes variable dependent:

$$|E, x_1, x_2, ...> = a_E{}^\dagger(x_1, x_2, ...)|0> \qquad (4.3.10)$$

as does the lowering operator:

$$a_E(x_1, x_2, ...)|E, x'_1, x'_2, ... > = \delta(x_1, x'_1)\,\delta(x_2, x'_2)...|0> $$
$$\qquad\qquad (4.3.11)$$

where $\dagger$ signifies the hermitean conjugate, and $\delta(x_k, x'_k)$ is a Kronecker-type δ for discrete variables and a Dirac-type δ-function for continuous variables.

Again we do not wish repeated application of a raising operator to generate states with higher eigenvalues. Therefore

$$a_E^\dagger(x_1, x_2, \ldots)a_E^\dagger(x_1, x_2, \ldots)|0> = 0 \qquad (4.3.12)$$
$$a_E(x_1, x_2, \ldots)a_E(x_1, x_2, \ldots)|E> = 0 \qquad (4.3.13)$$

etc. Consequently the reader can verify that the raising and lowering operators must satisfy the anicommutation relations:

$$\{a_E^\dagger(x_1, x_2, \ldots), a_E(x'_1, x'_2, \ldots)\} = gn(E)\delta(x_1, x'_1)\delta(x_2, x'_2)\ldots$$
$$\{a_E(x_1, x_2, \ldots), a_E(x'_1, x'_2, \ldots)\} = 0 \qquad (4.3.14)$$
$$\{a_E^\dagger(x_1, x_2, \ldots), a_E^\dagger(x'_1, x'_2, \ldots)\} = 0$$

More generally if there are a number of semantic eigenvalue operators A_{E_k} where $k = 1, 2, \ldots, q$ then the anticommutation relations are

$$\{a_{E_k}^\dagger(x_1, x_2, \ldots), a_{E_k}(x'_1, x'_2, \ldots)\} = gn(E)\delta_{kk'}\delta(x_1, x'_1)\delta(x_2, x'_2)\ldots$$
$$\{a_{E_k}(x_1, x_2, \ldots), a_{E_k}(x'_1, x'_2, \ldots)\} = 0 \qquad (4.3.15)$$
$$\{a_{E_k}^\dagger(x_1, x_2, \ldots), a_{E_{k'}}^\dagger(x'_1, x'_2, \ldots)\} = 0$$

where $\delta_{kk'}$ is a Kronecker delta. The semantic eigenvalue operators can be represented in terms of thse operators by

$$A_{E_k}(x_1, x_2, \ldots) = a_{E_k}^\dagger(x_1, x_2, \ldots)a_{E_k}(x_1, x_2, \ldots) \qquad (4.3.16)$$

and the projection operators are

$$P_{E_k}(x_1, x_2, \ldots) = P'_{E_k}(x_1, x_2, \ldots)\tfrac{1}{2}[1 + [gn(E_k)]^{-1}a_{E_k}^\dagger(x_1, x_2, \ldots)a_{E_k}(x_1, x_2, \ldots)]$$
$$(4.3.17)$$

The calculus eigenvalue operators, and raising and lowering operators have a similar form

$$\{b_{E_k}^\dagger(x_1, x_2, \ldots), b_{E_k}(x'_1, x'_2, \ldots)\} = gn(E_k)\delta_{kk'}\delta(x_1, x'_1)\delta(x_2, x'_2)\ldots$$
$$\{b_{E_k}(x_1, x_2, \ldots), b_{E_k}(x'_1, x'_2, \ldots)\} = 0 \qquad (4.3.18)$$

$$\{b_{E_k}^{\dagger}(x_1, x_2, \ldots), b_{E_{k'}}^{\dagger}(x'_1, x'_2, \ldots)\} = 0$$

with

$$B_{E_k}(x_1, x_2, \ldots) = b_{E_k}^{\dagger}(x_1, x_2, \ldots)b_{E_k}(x_1, x_2, \ldots) \qquad (4.3.19)$$

and

$$Q_{E_k}(x_1, x_2, \ldots) = \tfrac{1}{2}[1 + [gn(E_k)]^{-1}b_{E_k}^{\dagger}(x_1, x_2, \ldots)b_{E_k}(x_1, x_2, \ldots)]Q'_{E_k}(x_1, x_2, \ldots) \qquad (4.3.20)$$

4.3.3 Effect of Raising and Lowering Operators on Universes of Discourse

Consider the effect of a lowering operator on a universe state such as

$$a_{E_k}|E_1, E_2, E_3, \ldots, E_k, \ldots, E_n> = |E_1, E_2, E_3, \ldots, \overset{k^{th}}{0}, \ldots, E_n>$$

$$(4.3.21)$$

The kth eigenvalue becomes zero, which is the Gödel number of the empty string of characters (symbols).

Now consider a semantic universe of discourse (Parallel observations and comments can be made for a calculus universe of discourse.) with terms corresponding to Gödel numbers: $E_1, E_2, \ldots, E_n$.[76] The state $|\Omega_u>$ corresponding to this universe of discourse is given by eq. 3.4.3.

A universe of discourse state $|\Omega_u>$ can be "restricted" using products of lowering operators. For example

$$a_{E_k}|\Omega_u> = gn(E_k)|\Omega'_u> \qquad (4.3.22)$$

where the kth eigenvalue of $|\Omega'_u>$ is 0 and

$$|\Omega'_u> = |E_1, E_2, E_3, \ldots, \overset{k^{th}}{0}, \ldots, E_n> \qquad (4.3.23)$$

The axioms, theorems and proofs of the universe of discourse state $|\Omega'_u>$ are those of the universe of discourse $|\Omega_u>$ with the kth term

---

[76] Zero is the Gödel number of the empty string.

being zero. The kth term then corresponds to the empty string when mapped to English.

Thus a statement such as

$$<\Omega'_u| \sum_{n \in \{m'\}} A_{E_n}|\Omega'_u> \qquad (4.3.24a)$$

satisfies the corresponding projection operator product equation

$$F(P'_{E_1}, P'_{E_2}, ...) \prod_{n \in \{m\}} P_{E_n}|\Omega'_u> = 0 \qquad (4.3.24b)$$

where $F(P'_{E_1}, P'_{E_2}, ...)$ is a function of P'_{E_1}, P'_{E_2} and so on, if k is among the set of integers $\{m\}$ and thus the expectation value[77]

$$<\Omega'_u|F(P'_{E_1}, P'_{E_2}, ...) \prod_{n \in \{m\}} P_{E_n}|\Omega'_u> = 0 \qquad (4.3.25)$$

for each statement in the universe of discourse $|\Omega_u>$ containing the kth term. Thus the set of statements with a non-zero truth value in the $|\Omega'_u>$ universe of discourse consists of those statements not containing the kth term. $|\Omega'_u>$ is properly regarded as a subuniverse of discourse.

Whether the subuniverse of discourse is of interest is another question. In the next chapter we consider a simple case where the generation of subuniverses is of interest.

To summarize: A statement in a subuniverse defined above has the form

$$<\Omega'_u| \sum_{n \in \{m'\}} A_{E_n}|\Omega'_u> \qquad (4.3.26)$$

for the subset of the integers of $\{m\}$, namely $\{m'\}$ not containing k.

The projection operator product corresponding to this statement has the form

$$F(P'_{E_1}, P'_{E_2}, ...) \prod_{n \in \{m'\}} P_{E_n} \qquad (4.3.27)$$

---

[77] Inner product.

and whether a statement in the subuniverse has a truth value is determined by the expectation value

$$<\Omega'_u| \ F(P'_{E_1}, P'_{E_2}, \ldots)\prod_{n \in \{m'\}} P_{E_n}|\Omega'_u> \qquad (4.3.28)$$

If it is zero then it has no truth value; if it is non-zero then it has a truth value. The statements of this subuniverse exclude statements of the parent universe containing the kth term.

We now have a method for generating subuniverses through the application of one or more lowering operators. We shall see examples of subuniverses in the next chapter.

4.3.4 Multiple Lowering Operator Case

In the case where we apply q lowering operators (whose indices form the set $\{p\}$) to a universe state[78], then eq. 4.3.22 generalizes directly to

$$\prod_{n \in \{p\}} a_{E_n}|\Omega_u> \ = \ \prod_{n \in \{p\}} gn(E_n)|\Omega''_u> \qquad (4.3.29)$$

and members of the set of statements with truth values in this subuniverse have the form

$$<\Omega''_u| \ \sum_{n \in \{m-p\}} A_{E_n}|\Omega''_u> \qquad (4.3.30)$$

where $\{m - p\}$ is the set of integers in $\{m\}$ that does not include any integers in $\{p\}$.

The corresponding projection operator product expectation value has the form

$$<\Omega''_u| \ F(P'_{E_1}, P'_{E_2}, \ldots) \prod_{n \in \{m-p\}} P_{E_n}|\Omega''_u> \qquad (4.3.31)$$

The truth value of the statement is determined by the value of eq. 4.3.31: if the value is zero then the corresponding statement has no truth value; if the value is non-zero then the corresponding statement has a truth value.

---

[78] $|\Omega_u>$ is defined by eq. 3.4.3 and $\{m\}$ is defined in the statement following it.

5. Hierarchies of Universes of Discourse

In this chapter we will define hierarchies of subuniverses and consider some simple examples. In addition we will show how to combine universes into "superuniverses" – the equivalent of what we normally call unified theories.

5.1 Hierarchies of Subuniverses

In the previous chapter we examined the case of a two level hierarchy. One can construct multi-level hierarchies of universes using various sets of lowering operator products at each level. Fig. 5.1 shows a three level hierarchy diagramatically. The procedure is so simple that writing the explicit equations generating the hierarchy is unnecessary in the author's view. They are similar to those given for a two level hierarchy in section 4.3.4.

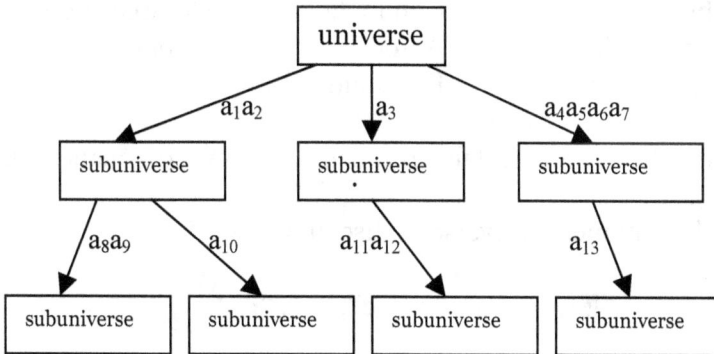

Figure 5.1. A hierarchy of universes with the lowering operators indicated for the transition to each subuniverse.

One primary purpose in defining subuniverse hierarchies is to enable separate universes of discourse that have a significant overlap in their primitive terms, axioms and theorems to be united in a superuniverse through the reversal of the mechanism described for creating subuniverse hierarchies. We form a unified set of terms with one subset containing the primitive terms of one universe and another subset containing the primitive terms of the other universe. Then a composite set of axioms is constructed such that setting one set of terms to zero (empty strings) yields one of the initial universes and setting a different set of terms to zero yields the other universe. Thus one achieves a unification of what had been two separate but similar universes of discourse.[79] The arithmetic universe considered next is an example of this process.

5.2 Subuniverses of a Simple Arithmetical Universe of Discourse

We will consider a simple universe of discourse, that is similar to Gödel's "formal system P"[80], to illustrate the utility of creating subuniverses of discourse and to illustrate combining universes to produce a super universe. The axioms in this universe are:

1. $a + b = b + a$ Commutativity of Addition
2. $a + (b + c) = (a + b) + c$ Associativity of Addition
3. $a{\cdot}b = b{\cdot}a$ Commutativity of Multiplication
4. $a{\cdot}(b{\cdot}c) = (a{\cdot}b){\cdot}c$ Associativity of Multiplication
5. $a{\cdot}(b + c) = a{\cdot}b + a{\cdot}c$ Distributive Law

They are the fundamental arithmetic axioms of natural numbers (integers).

The arithmetic universe of discourse state $|\Omega_u>$ is:

$$|\Omega_u> = |\text{"a", "b", "c", "+", ".", "=", "(", ")"} > \qquad (5.1)$$

---

[79] J. C. Maxwell achieved this feat by successfully uniting electricity and magnetism in his theory of Electromagnetism in 1865.
[80] Gödel (1962) p. 41.

5.2.1 Additive Universe of Discourse

If we now apply the lowering operator for multiplication "·", namely $a_{\cdot,\cdot}$, to $|\Omega_u>$ then the universe of discourse is reduced to addition only:

1. $a + b = b + a$ Commutativity of Addition
2. $a + (b + c) = (a + b) + c$ Associativity of Addition

with the subuniverse state:

$$|\Omega_+> = |"a", "b", "c", "+", 0, "=", "(", ")" > \qquad (5.2)$$

All statements containing a multiplication become statements without a truth value when their projection operator products are evaluated between $<\Omega_+|$ and $|\Omega_+>$. We call these statements *null statements* since they are neither true nor false. (They could also be viewed as undecidable.)

5.2.2 Multiplicative Universe of Discourse

The multiplicative universe of discourse state $|\Omega_\bullet>$ is obtained by applying the lowering operator for addition "+", namely $a_{\cdot,+,\cdot}$, to $|\Omega_u>$. The subuniverse state $|\Omega_\bullet>$ is:

$$|\Omega_\bullet> = |"a", "b", "c", 0, "·", "=", "(", ")" > \qquad (5.3)$$

The axioms in this subuniverse are the purely multiplicative axioms:

1. $a·b = b·a$ Commutativity of Multiplication
2. $a·(b·c) = (a·b)·c$ Associativity of Multiplication

Thus we obtain the multiplicative subuniverse. All statements containing an addition become null statements when their projection operator products are evaluated between $<\Omega_\bullet|$ and $|\Omega_\bullet>$.

5.3 Subuniverses of Euclidean Geometry

Euclidean geometry was plagued for millenia by concerns over the fifth postulate. In the 19th Century Bolyai and Lobachevsky showed that omitting the fifth postulate opened the door to non-Euclidean geometries that were valid on certain curved surfaces. Euclidean geometry has five axioms plus five common notions including the notion of congruence. The primitive terms and axioms are:

Primitive terms:
point, line, straight line, circle, angle, congruent (in common notions)

Axioms:
1. Given two points there is a straight line that joins them.
2. Any straight line can be prolonged indefinitely.
3. A circle can be constructed when its center, and a point on it, are specified.
4. All right angles are equal.
5. If a straight line falling on two straight lines makes the interior angles on the same side sum to less than two right angles, the two straight lines if prolonged indefinitely, meet on that side on which the angles are less than two right angles.

They also include some common English words.

If we consider the semantic universe of discourse of Euclidean geometry, then the interpreted primitive terms, the axioms, and the common notions comprise the knowledge basis of the Euclidean universe of discourse.

If we define a semantic subuniverse by setting "congruent" to the empty string using a lowering operator, and reinterpreting the primitive term "straight line" so that axiom 5 is not necessarily true, then we obtain a family of geometries that includes Euclidean geometry and non-Euclidean geometries.

Thus showing a subuniverse can be more general then its parent universe.

5.4 A Subuniverse of Electromagnetism

A more general form of the Maxwell equations (We will call it the pseudo-Maxwell equations.) in the absence of sources and currents can be defined by

$$\nabla \times \mathbf{E} = -\beta \partial \mathbf{B}/\partial t$$
$$\nabla \cdot \mathbf{E} = 0 \tag{5.4}$$

$$\nabla \cdot \mathbf{B} = 0$$
$$\nabla \times \mathbf{B} = - \beta \partial \mathbf{E}/\partial t$$

where β is a constant and $\times$ represents the vector cross product while $\cdot$ represents a vector inner product. If $\beta = 1$ then eqs. 5.4 are exactly the Maxwell equations of electrodynamics in the absence of sources and currents.

We now define the nine primitive terms of eqs. 5.4:

"$\nabla \times$"   "$\nabla \cdot$"   "="   "$-$"   "β"   "$\partial/\partial t$"   "$\mathbf{E}$"   "$\mathbf{B}$"   "0"

which we associate with the eigenvalue operators A_1, A_2, ... , A_9. The universe state is

$$|\Omega_u\rangle = |\text{ "}\nabla\times\text{", "}\nabla\cdot\text{", "="}, \text{ "}-\text{", "}\beta\text{", "}\partial/\partial t\text{", "}\mathbf{E}\text{", "}\mathbf{B}\text{", "0"}\rangle \qquad (5.5)$$

We apply the lowering operator to make the eigenvalue of A_5 zero producing the resulting universe state $|\Omega_{u1}\rangle$:

$$|\Omega_{u1}\rangle = c_{\text{"}\beta\text{"}}|\Omega_u\rangle$$

$$= |\text{ "}\nabla\times\text{", "}\nabla\cdot\text{", "="}, \text{ "}-\text{", } 0, \text{ "}\partial/\partial t\text{", "}\mathbf{E}\text{", "}\mathbf{B}\text{", "0"}\rangle \qquad (5.6)$$

Thus the first and fourth Maxwell equations become:

$$\langle\Omega_{u1}|(A_1 + A_7 + A_3 + A_9)|\Omega_{u1}\rangle \equiv \nabla \times \mathbf{E} = 0$$
$$\qquad (5.7)$$
$$\langle\Omega_{u1}|(A_1 + A_8 + A_3 + A_9)|\Omega_{u1}\rangle \equiv \nabla \times \mathbf{B} = 0$$

The complete set of the subuniverse "Maxwell" equations is

$$\nabla \times \mathbf{E} = 0$$
$$\nabla \cdot \mathbf{E} = 0$$
$$\nabla \cdot \mathbf{B} = 0 \qquad (5.8)$$
$$\nabla \times \mathbf{B} = 0$$

Eqs. 5.8 represent a "new" theory in which the electric and magnetic fields are decoupled and become separate "subtheories." An important result of this theory is the absence of electromagnetic waves.

The creation of subuniverses, and the reverse process of combining subuniverses to produce a superuniverse, provide a mechanism for relating theories, and a mechanism for unifying similar theories into a joint theory in some area of physics or mathematics.

5.5 Products of Universes

All universes, as we have defined them, are Hilbert spaces. Thus it is natural to consider direct products of universes. We will start by considering the case of the direct product of the additive and multiplicative subuniverses defined earlier in this chapter, namely $|\Omega_+\rangle|\Omega_\bullet\rangle$. Then the axioms

1. $a + b = b + a$	Commutativity of Addition
2. $a + (b + c) = (a + b) + c$	Associativity of Addition
3. $a \cdot b = b \cdot a$	Commutativity of Multiplication
4. $a \cdot (b \cdot c) = (a \cdot b) \cdot c$	Associativity of Multiplication

emerge directly for the product of subuniverses. For example, axiom 1 is generated from

$$\langle\Omega_\bullet|\langle\Omega_+|(A_{+"a"}I_\bullet + A_{+"+"}I_\bullet + A_{+"b"}I_\bullet + A_{+"="}I_\bullet + A_{+"b"}I_\bullet + A_{+"+"}I_\bullet + A_{+"a"}I_\bullet)|\Omega_+\rangle|\Omega_\bullet\rangle$$

where $I_\bullet$ is the identity operator in the multiplicative subuniverse. The other axioms are similarly generated. The plus subscript on the operators A_+ indicates they are operators in the additive subuniverse. Since we are taking the direct product of Hilbert spaces the eigenvalue operators are paired from each space as shown.

The fifth axiom

5. $a \cdot (b + c) = a \cdot b + a \cdot c$	Distributive Law

which combines addition and multiplication can also be added to the set of axioms. The statement of the fifth axiom is slightly more complex since both subuniverses "interact" through this axiom:

$$\langle\Omega_\bullet|\langle\Omega_+|(A_{+"a"}I_\bullet + A_{\bullet"''}I_+ + A_{+"c"}I_\bullet + A_{+"b"}I_\bullet + A_{+"+"}I_\bullet + A_{+"c"}I_\bullet + A_{+"y"}I_\bullet + A_{+"="}I_\bullet +$$
$$+ A_{+"a"}I_+ + A_{\bullet"''}I_+ + A_{+"b"}I_\bullet + A_{+"+"}I_\bullet + A_{+"a"}I_\bullet + A_{\bullet"''}I_+ + A_{+"c"}I_\bullet)|\Omega_+\rangle|\Omega_\bullet\rangle$$

The "$\bullet$" subscript on the operators $A_\bullet$ indicates they are operators in the multiplicative subuniverse. The axiom can be restated in alternate ways as well since we identify the terms "a", "b", and "c" in $|\Omega_+\rangle$ with the corresponding terms in $|\Omega_\bullet\rangle$.

This example illustrates direct products of universes. It also shows that an issue, that may arise, is duplicate terms in the universes. As shown this issue is usually easily resolved.

The procedure of taking the direct product of two universes can result in the creation of a "superuniverse" that combines the features of both universes. Additional axioms can be added to the combined universe to provide inter-relationships between the universes. Axiom 5 above is an example of an axiom that interrelates (causes an interaction between) the component universes. Thus the combination of universes can lead to more than their initial combined content.

6. Matrix Form of Operator Logic

The eigenvalue operators A_{E_k} and B_{E_k} in the semantic and calculus universes of discourse each have two eigenvalues: a Gödel number $gn(E_k)$ and zero (which is the Gödel number of the empty string). For each k, the operator A_{E_k} or B_{E_k}, and each operator's two eigenstates $|0>$ and $|E_k>$, can be expressed in terms of SU(2) operators and eigenstates in a 2×2 matrix formulation. This representation is also used to describe spin ½ particles such as electrons and quarks in physics theories. Later when we develop the Platonic realm of Ideas and their relation to Reality this correspondence will turn out to be of importance.

6.1 Spinor Representation of Operator Logic

The SU(2) spinor representation is based on the Pauli matrices:

$$\sigma_1 = \begin{bmatrix} 0 & 1 \\ 1 & 0 \end{bmatrix} \qquad \sigma_2 = \begin{bmatrix} 0 & -i \\ i & 0 \end{bmatrix} \qquad \sigma_3 = \begin{bmatrix} 1 & 0 \\ 0 & -1 \end{bmatrix} \qquad (6.1)$$

The below matrices in this section can be simply expressed in the spinor representation. Consider each eigenvalue operator A_{E_k} for some value k.[81] We can represent this operator and its eigenvectors as a matrix and column vectors:

$$A_{E_k} \rightarrow \begin{bmatrix} gn(E_k) & 0 \\ 0 & 0 \end{bmatrix} \qquad |E_k> \rightarrow \begin{bmatrix} 1 \\ 0 \end{bmatrix} = s_{k\uparrow} \qquad |0> \rightarrow \begin{bmatrix} 0 \\ 1 \end{bmatrix} = s_{k\downarrow}$$

$$(6.2)$$

---

[81] Calculus Logic Operators have a completely analogous formalism.

The raising and lowering operators, which are used in the spinor representation, are

$$c_{E_k}{}^\dagger = \begin{bmatrix} 0 & 1 \\ 0 & 0 \end{bmatrix} \qquad c_{E_k} = \begin{bmatrix} 0 & 0 \\ 1 & 0 \end{bmatrix} \tag{6.3}$$

where † signifies hermitean conjugation, and where

$$c_{E_k}{}^\dagger \equiv [gn(E_k)]^{-\frac{1}{2}} a_{E_k}{}^\dagger \tag{6.4}$$

$$c_{E_k} \equiv [gn(E_k)]^{-\frac{1}{2}} a_{E_k}$$

with the anticommutation relations

$$c_{E_k}{}^\dagger c_{E_{k'}} + c_{E_{k'}} c_{E_k}{}^\dagger \equiv \{c_{E_k}{}^\dagger, c_{E_{k'}}\} = \delta_{kk'}$$
$$\{c_{E_k}, c_{E_{k'}}\} = 0 \tag{6.5}$$
$$\{c_{E_k}{}^\dagger, c_{E_{k'}}{}^\dagger\} = 0$$

where $\delta_{kk'}$ is the Kronecker delta. Note the semantic eigenvalue operator A_{E_k} then becomes

$$A_{E_k} = gn(E_k)c_{E_k}{}^\dagger c_{E_k} \tag{6.6}$$

and the corresponding projection operator is

$$P_{E_k} = P'_{E_k}\tfrac{1}{2}[1 + c_{E_k}{}^\dagger c_{E_k}] \tag{6.7}$$

where P'_{E_k} is a truth projection operator. (See section 3.8.1 for details on truth projection operators.)

The universe of discourse $|\Omega_u\rangle$ with q factors is generated by the direct product of the q individual eigenstates $s_{i\uparrow}$ of each primitive term in the universe:

$$|\Omega_u\rangle = \prod_i s_{i\uparrow} \tag{6.8a}$$

Note the universe state is normalized to one, $\langle\Omega_u|\Omega_u\rangle = 1$. A statement has the form:

$$\text{statement} = \langle\Omega_u|A_1 + A_2 + A_3 + \dots |\Omega_u\rangle \qquad (6.8b)$$

where the matrices A_i have the form specified in eq. 6.2. The corresponding truth value is given by

$$\text{Truth Value} = \langle\Omega_u| F(P'_1, P'_2, \dots)P_1 P_2 P_3 \dots |\Omega_u\rangle \qquad (6.8c)$$

A simple example of this formalism is a two term universe. In this universe of discourse

$$|\Omega_u\rangle = s_{1\uparrow} s_{2\uparrow}$$

The statement
$$\text{statement} = \langle\Omega_u|A_1 + A_2|\Omega_u\rangle$$
$$= gn(E_1) + gn(E_2)$$

can be transformed into a human language by mapping the Gödel numbers to their human language equivalent, dropping the + sign, and putting the statement into a grammatically correct form for that language. This might include add "a" and "the" (in English), and possibly reordering the words. (For example in German words often appear in a different order than English.)

6.2 A Symmetry of the Spinor Matrix Representation

Eq. 6.2 implies a discrete symmetry is present in the matrix form of a universe of discourse[82]. If we define the eigenvalue and projection operators

$$A''_{E_k} = \sigma_1^{-1}A_{E_k}\sigma_1 \qquad (6.9a)$$
$$P''_{E_k} = \sigma_1^{-1}P_{E_k}\sigma_1 \qquad (6.9b)$$
$$P'''_{E_k} = \sigma_1^{-1}P'_{E_k}\sigma_1 \qquad (6.9c)$$

for all k using $\sigma_1^{-1} = \sigma_1$,[83] and if we define a new universe state

---

[82] Similar comments apply to the matrix form of a calculus universe of discourse.

$$|\Omega_u> = \sigma_1^{-1}|\Omega''_u>$$ (6.10)

then the transformed statements and their truth values in the new matrix representation are the same as that of the original matrix representation.

A statement and its truth value (eqs. 3.4.1 and 3.4.2) is transformed to

statement $= <\Omega_u|A_1 + A_2 + A_3 \ldots |\Omega_u> = <\Omega''_u|A''_1 + A''_2 + \ldots |\Omega''_u>$
(6.11)

and

$$TV = \text{Truth Value} = <\Omega_u|F(P'_{E_1}, P'_{E_2}, \ldots)\prod_{n \in \{m'\}} P_{E_n} |\Omega_u>$$

$$= <\Omega''_u|F(P'''_{E_1}, P'''_{E_2}, \ldots)\prod_{n \in \{m'\}} P''_{E_n} |\Omega''_u>$$ (6.12)

Thus we have invariance under the discrete transformation σ_1.

6.3 Direct Sum of Spinor Representations of Universes – Dirac Matrices

In this section we will consider the direct sum of the spinor representations of two universes that are not necessarily the same. In part, our goal is to construct a space whose eigenvectors are similar to 4-component Dirac spinors. We will use these eigenvectors later to develop the relation of the Platonic realm of Ideas to Reality.

The universes that we will define may have different terms, axioms or proof schema. We will first define the direct sum and then introduce terms, which intermix the two universes of discourse to create an interrelated universe. We denote the constructs of the first universe as

$$A_{E1k} \qquad P_{E1k} \qquad c_{E1k} \qquad c_{E1k}^\dagger \qquad s_{1k\uparrow} \qquad s_{1k\downarrow} \qquad |\Omega_{1u}> \quad (6.13)$$

and the constructs of the second universe as

---

[83] There is, in principle, a σ_1 for each value of k. They, of course, all have identically the same form.

$$A_{E2k} \quad P_{E2k} \quad c_{E2k} \quad c_{E2k}^{\dagger} \quad S_{2k\uparrow} \quad S_{2k\downarrow} \quad |\Omega_{2u}\rangle \quad (6.14)$$

From these constructs we form four column vectors with four columns:

$$S_{\uparrow\uparrow k} = \begin{bmatrix} S_{1k\uparrow} \\ S_{2k\uparrow} \end{bmatrix} \quad S_{\uparrow\downarrow k} = \begin{bmatrix} S_{1k\uparrow} \\ S_{2k\downarrow} \end{bmatrix} \quad S_{\downarrow\uparrow k} = \begin{bmatrix} S_{1k\downarrow} \\ S_{2k\uparrow} \end{bmatrix} \quad S_{\downarrow\downarrow k} = \begin{bmatrix} S_{1k\downarrow} \\ S_{2k\downarrow} \end{bmatrix} \quad (6.15)$$

and[84] 4×4 matrices composed of 2×2 matrix entries

$$A_{E_k} = \begin{bmatrix} A_{E1k} & 0 \\ 0 & A_{E2k} \end{bmatrix} \quad P_{E_k} = \begin{bmatrix} P_{E1k} & 0 \\ 0 & P_{E2k} \end{bmatrix} \quad (6.16)$$

$$c_{E_k} = \begin{bmatrix} c_{E1k} & 0 \\ 0 & c_{E2k} \end{bmatrix} \quad d_{E_k} = \begin{bmatrix} c_{E1k} & 0 \\ 0 & c_{E2k}^{\dagger} \end{bmatrix} \quad (6.17)$$

and also $c_{E_k}^{\dagger}$ and $d_{E_k}^{\dagger}$. The direct sum universe state is

$$|\Omega_u\rangle = \begin{bmatrix} |\Omega_{1u}\rangle \\ |\Omega_{2u}\rangle \end{bmatrix} \quad (6.18)$$

6.3.1 Independent Universes

We can express all the statements in the first universe by projecting out the parts associated with the second universe. The universe projection operator that plays this role is

---

[84] If the number of terms in universe one and universe two differ then one pads eq. 6.16 with zeroes for the "missing" eigenvalue operators.

$$U_1 = \tfrac{1}{2}(I_4 + \gamma^0) \tag{6.19}$$

where I_4 is the 4×4 identity matrix and γ^0 is one of the four Dirac γ matrices.[85] The four Dirac γ matrices and the product $\gamma^5 = \gamma_5 = i\gamma^0\gamma^1\gamma^2\gamma^3$ are:

$$\gamma^0 = \begin{bmatrix} I & 0 \\ 0 & -I \end{bmatrix} \quad \gamma^i = \begin{bmatrix} 0 & \sigma_i \\ -\sigma_i & 0 \end{bmatrix} \quad \gamma^5 = \begin{bmatrix} 0 & I \\ I & 0 \end{bmatrix}$$

$$\tag{6.20}$$

for $i = 1, 2, 3$ where the submatrices are the 2×2 Pauli matrices and the 2×2 identity matrix I. We will use the γ^5 matrix in the next subsection.

Based on eqs. 3.4.1 we see that a statement strictly of the first universe has the form

$$\text{Universe 1 statement} = <\Omega_u|U_1(A_1 + A_2 + A_3 \dots)|\Omega_u> \tag{6.21}$$

where the eigenvalue operators A_i have the form of eq. 6.16 and its corresponding truth value is given by

$$\text{Universe 1 Truth Value} = <\Omega_u|U_1F(P'_1, P'_2, \dots)P_1 P_2 P_3 \dots|\Omega_u> \tag{6.22}$$

where the projection operators P_i have the form of eq. 6.16.

Similarly if we define the universe projection operator

$$U_2 = \tfrac{1}{2}(I_4 - \gamma^0) \tag{6.23}$$

then we see that a statement strictly of the second universe has the form

$$\text{Universe 2 statement} = <\Omega_u|U_2(A_1 + A_2 + A_3 \dots)|\Omega_u> \tag{6.24}$$

where the eigenvalue operators A_i have the form of eq. 6.16 and its corresponding truth value is given by

---

[85] We introduce Dirac γ matrices with a Platonic view of connecting Operator Logic to Elementary Particle Physics later in this volume.

$$\text{Universe 2 Truth Value} = <\Omega_u|U_2 F(P'_1, P'_2, \ldots)P_1 P_2 P_3 \ldots|\Omega_u> \quad (6.25)$$

where the projection operators P_i have the form of eq. 6.16. Thus both universes are embodied in the direct sum.

6.3.2 Mixing Between the Universes

In the preceeding subsection both universes in the direct sum maintained their distinct identity with separate sets of statements. In this section we will consider statements that have terms from both universes. This possibility is of interest because it is analogous to the mixing of two component particle spinors in the Dirac equation in Physics that is caused by the particle's mass term. Again we are looking to the development of the Platonic Ideas – Reality relationship later. This possibility of mixing universes is of interest in its own right as well.

In order to create mixed statements with terms that come from both universes we have to define eigenvalue and projection matrices that perform that task:[86]

$$\hat{A}_{E_k} = \gamma^5 A_{E_k} \qquad \text{and} \qquad \hat{S} = \gamma^5 P_{E_k} \qquad (6.26)$$

using the matrices in eq. 6.16. The new matrices have the form

$$\hat{A}_{E_k} = \begin{bmatrix} 0 & A_{E_{2k}} \\ A_{E_{1k}} & 0 \end{bmatrix} \qquad \hat{S}_{E_k} = \begin{bmatrix} 0 & P_{E_{2k}} \\ P_{E_{1k}} & 0 \end{bmatrix} \qquad (6.27)$$

Some examples of statements that mix both universes are

$$\text{Mixed Universe statement} = <\Omega_u|U_1(A_1 + \ldots + \hat{A}_i + \ldots + A_j + \ldots + \hat{A}_k + \ldots)|\Omega_u> \quad (6.28)$$

and

$$\text{Mixed Universe statement} = <\Omega_u|U_2(A_1 + \ldots + \hat{A}_i + \ldots + A_j + \ldots + \hat{A}_k + \ldots)|\Omega_u> \quad (6.29)$$

---

[86] There is also an alternate approach based on the use of a mass term, which we will discuss later.

and their corresponding truth values have the form

Mixed Universe Truth Value =
$$= <\Omega_u|U_1F(P'_1, P'_2, ..., S'_1, S'_2, ...)P_1 ... \hat{S}_i ... P_j ... \hat{S}_k ... |\Omega_u>$$
(6.30)

and

Mixed Universe Truth Value =
$$= <\Omega_u|U_2F(P'_1, P'_2, ..., S'_1, S'_2, ...)P_1 ... \hat{S}_i ... P_j ... \hat{S}_k ... |\Omega_u>$$
(6.31)

respectively.

A reader who is a logician may wonder why we have developed the concept of mixed universe statements. Again we note this concept is analogous to the Dirac equation for spin ½ elementary particles and will be used later as we attempt to construct a realization of Plato's theory of Ideas and Reality. The analogy is also evident in the two component form of the Dirac equation. The Dirac wave function ψ has four components that can be viewed as a pair of spinors:

$$\psi = \begin{bmatrix} \psi_+ \\ \psi_- \end{bmatrix}$$
(6.32)

that satisfy the two-component form of the Dirac equation:

$$i\partial\psi_+/\partial t = -i\boldsymbol{\sigma}\cdot\nabla\psi_+ - m\psi_-$$
(6.33)
$$i\partial\psi_-/\partial t = i\boldsymbol{\sigma}\cdot\nabla\psi_- - m\psi_+$$
(6.34)

where m is the particle's mass, $\boldsymbol{\sigma}$ is a 3-vector whose components are he Pauli matrices, and ∇ is the grad differential operator. If we view these equations as statements, which is what they are, then the mass terms mix the "universe" of ψ_+ spinors with the "universe" of ψ_- spinors. Thus the analogy. We will discuss this in more detail at a later point.[87]

---

[87] Eqs. 6.33 and 6.34 also show we could create a spinor representation for spin ½ particles directly. Therefore forming the direct sum (eqs. 6.14 – 6.18) was not necessary but it was helpful

The fact that γ_5 appears in the Operator Logic mixing terms (eq. 6.26) is significant also since γ_5 also appears in the Dirac equation formalism for left-handed and right-handed neutrinos. (Neutrinos are also spin ½ particles.)

6.4 Vector Operator Logic

In this section we will define a vector matrix representation of Operator Logic. It is possible that statements in a universe of discourse have terms that appear in logically related pairs. It is also not uncommon that a universe of discourse (or a deductive system) consists of a number of distinct sectors that are intertwined by the addition of one or more axioms that reference terms and operations in other sectors. Section 5.2 provides an example of this situation with two axioms for addition only, two axioms for multiplication only, and a third axiom that combines addition and multiplication operations.

The most important reason for considering Vector Operator Logic is our desire to build the Platonic connection between Ideas and Reality. The elementary particles, with which we are familiar, are spin ½ fermions and spin 1 vector bosons (plus spin 2 gravitons). Thus Vector Operator Logic representations may be relevant to connect Ideas with Reality. (We will forego presenting the spin 2 Operator Logic representation in this book.)

We have seen the spinor matrix formulation (section 6.1) of universes of discourse earlier. It is also possible to use a vector formalism for Operator Logic that might be more appropriate for some types of universes. The matrix Vector Operator Logic formalism that we will consider in this section is similar to the formalism for spin 1 physical systems.[88]

We will use the 3-dimensional (vector) representation of SU(2). The three generator matrices, T_i, of SU(2) can be placed in a form where T_3 is diagonal:

---

because it led directly to a Dirac matrix formulation that is conventional in spin ½ particle physics.

[88] It is also possible to develop spin 3/2, spin 2, and higher spin formalisms for universes of discourse. Their development parallels the spin ½ and spin 1 (vector) cases discussed here.

$$T_3 = \begin{bmatrix} 1 & 0 & 0 \\ 0 & 0 & 0 \\ 0 & 0 & -1 \end{bmatrix} \quad T_+ = \begin{bmatrix} 0 & \sqrt{2} & 0 \\ 0 & 0 & \sqrt{2} \\ 0 & 0 & 0 \end{bmatrix} \quad T_- = \begin{bmatrix} 0 & 0 & 0 \\ \sqrt{2} & 0 & 0 \\ 0 & \sqrt{2} & 0 \end{bmatrix}$$

$$(6.35)$$

The kth eigenvalue and projection operators can have the form:

$$A_{E_k} = \begin{bmatrix} gn(E_{k1}) & 0 & 0 \\ 0 & gn(E_{k2}) & 0 \\ 0 & 0 & 0 \end{bmatrix} \quad P_{E_k} = \begin{bmatrix} P_{Ek1} & 0 & 0 \\ 0 & P_{Ek2} & 0 \\ 0 & 0 & 0 \end{bmatrix} \quad (6.36)$$

where A_{Ek1} and A_{Ek2} are eigenvalue operators for the specific Gödel eigenvalues $gn(E_{k1})$ and $gn(E_{k2})$ respectively that are equivalent to specific terms of the universe, and P_{Ek1} and P_{Ek2} are their corresponding projection operators.

For each integer value k, the three eigenvectors are

$$s_{k1} = \begin{bmatrix} 1 \\ 0 \\ 0 \end{bmatrix} \quad s_{k2} = \begin{bmatrix} 0 \\ 1 \\ 0 \end{bmatrix} \quad s_{k3} = \begin{bmatrix} 0 \\ 0 \\ 1 \end{bmatrix} \quad (6.37)$$

Assuming there are q terms in the universe then an individual eigenstate will be a direct product with the form:

$$|\{m\}> = \prod_{\substack{k=1 \\ i \in \{m\}}}^{q} s_{ki} \quad (6.38)$$

where $\{m\}$ is an ordered set of values of $i = 1, 2, 3$. For example, if $q = 1$ and $\{m\} = \{1, 3, 1, 2, 2\}$ then

$$|\{m\}> = s_{11}s_{13}s_{11}s_{12}s_{12}$$

is an outer product of 5 eigenvectors.

The universe state $|\Omega_u>$, if there are q primitive terms, is the product of q eigenvectors:

$$|\Omega_u> = 2^{-q/2} \prod_{k=1}^{q} (s_{k1} + s_{k2})$$ (6.39)

The universal state is normalized to unity:[89]

$$<\Omega_u|\Omega_u> = 1$$ (6.40)

6.4.1 Vector Statements

Vector universes generate two statements if one does not mix the first and second rows of the eigenvalue operators. The "top" statement statement and its truth value have the forms

$$statement = <\Omega_u|U_1(A_i + A_j + A_k \ ... \ )|\Omega_u>$$ (6.41)

and

$$Truth \ Value = <\Omega_u| \ U_1F(P'_1, P'_2, ...)(P_i P_j P_k \ ... \ )|\Omega_u>$$ (6.42)

where U_1 is a projection operator defined below. The "second" statement statement and its truth value have the forms

$$statement = <\Omega_u|U_2(A_i + A_j + A_k \ ... \ )|\Omega_u>$$ (6.41a)

and

$$Truth \ Value = <\Omega_u|U_2F(P'_i, P'_j, ...)P_i P_j P_k \ ... \ |\Omega_u>$$ (6.42b)

The projection operators are:

$$U_1 = \begin{bmatrix} 1 & 0 & 0 \\ 0 & 0 & 0 \\ 0 & 0 & 0 \end{bmatrix} \qquad U_2 = \begin{bmatrix} 0 & 0 & 0 \\ 0 & 1 & 0 \\ 0 & 0 & 0 \end{bmatrix}$$ (6.43)

---

[89] Using the factor of $2^{-q/2}$.

6.4.2 S_3 Symmetry of Vector Universe of Discourse

The matrix formulation of Vector Operator Logic has a discrete S_3 symmetry[90] under permutations of row and column numbers. For example the interchange of row 1 and column 1 with row 2 and column 2 throughout does not change any statement. More generally if we define a symmetric group transformation S of the eigenvalue and projection operator matrices

$$A''_{E_k} = S^{-1}A_{E_k}S \qquad (6.44)$$
$$P''_{E_k} = S^{-1}P_{E_k}S \qquad (6.45)$$

for all k, and if we define a new universe state where S^{-1} is understood to be applied to each multiplicand in the direct products of eigenvectors:

$$|\Omega_u> = S^{-1}|\Omega''_u> \qquad (6.46)$$

then the transformed statements and their truth values in the new matrix representation are the same as that of the original matrix representation.

Under the symmetric group operation of interchanging rows and columns, 1 and 2, a statement and its truth value is transformed to

$$\text{statement} = <\Omega_u|U_1(A_1 + A_2 + A_3 \dots)|\Omega_u>$$
$$= <\Omega''_u|U_2(A''_1 + A''_2 + \dots)|\Omega''_u>$$
$$(6.47)$$

and

$$\text{Truth Value} = <\Omega_u|U_1(P_1 P_2 P_3 \dots)|\Omega_u>$$
$$= <\Omega''_u|U_2P''_1 P''_2 P''_3 \dots |\Omega''_u> \qquad (6.48)$$

Thus Vector Operator Logic has S_3 invariance.[91] The eigenvalue operators can be redefined to produce new operators that break S_3 symmetry.

---

[90] The symmetric group of degree 3 having six elements.
[91] The projections U_1 and U_2 must also be transformed under an S_3 transformation to maintain the invariance: $U_i'' = S^{-1}U_iS$ for i = 1, 2.

6.4.3 Raising and Lowering Operators to Define Vector Subuniverses

The raising and lowering operators that are used in the vector representation, are T_+ and T_- respectively. (See eq. 6.35). T_{k-} (the lowering operator for the kth term can be used to create a subuniverse with the kth terms shifted. Products of T_-'s can create subuniverses with several terms shifted. For example $T_{k-}T_{l-}T_{m-}T_{n-}$ shifts four terms: the kth, l^{th}, m^{th}, and n^{th} terms.

The shift action of the kth raising and lowering operator for the kth term's eigenvectors is

$$T_{k+}\ s_{k1} = 0$$
$$T_{k+}\ s_{k2} = \sqrt{2}\ s_{k1}$$
$$T_{k+}\ s_{k3} = \sqrt{2}\ s_{k2}$$

$$\hspace{6cm}(6.49)$$

$$T_{k-}\ s_{k1} = \sqrt{2}\ s_{k2}$$
$$T_{k-}\ s_{k2} = \sqrt{2}\ s_{k3}$$
$$T_{k-}\ s_{k3} = 0$$

where 0 represents the 3-vector composed of zeroes. We note

$$T_{k+}T_{k+}T_{k+} \equiv T_{k+}^{3} = 0 \hspace{2cm}(6.50)$$
$$T_{k-}T_{k-}T_{k-} \equiv T_{k-}^{3} = 0$$

We will consider an example of creating a subuniverse in the Vector Operator Logic. Applying T_{1-} to $|\Omega_u>$ yields

$$|\Omega'_u> = T_{1-}|\Omega_u> \hspace{2cm}(6.51)$$

with

$$<\Omega'_u|\Omega'_u> = 1 \hspace{2cm}(6.52)$$

The eigenvalue operator remains the same

$$A'_1 = A_1 \hspace{2cm}(6.53)$$

Thus the the statement

$$\text{statement} = <\Omega'_u|U_1(A'_1 + A_2)|\Omega'_u> \hspace{2cm}(6.54)$$

yields

$$\text{statement} = gn(E_{21}) \tag{6.55}$$

after evaluating the matrices A'_1 and A_2 between the universe state eq. 6.51.

If we define another subuniverse by

$$|\Omega''_u\rangle = T_2 T_{1-}|\Omega_u\rangle \tag{6.56}$$

noting

$$\langle\Omega''_u|\Omega''_u\rangle = 1 \tag{6.57}$$

and the new eigenvalue operators are

$$A''_1 = A_1 \tag{6.58}$$
$$A''_2 = A_2$$

then the statement

$$\text{statement} = \langle\Omega''_u|U_2(A''_1 + A''_2)|\Omega''_u\rangle \tag{6.59}$$

evaluates to[92]

$$\text{statement} = gn(E_{11}) + gn(E_{21}) \tag{6.60}$$

These examples illustrate the creation of subuniverses in Vector Operator Logic.

Vector Operator Logics may provide an appropriate formalism for some deductive systems. Spin 3/2, spin 2 and higher spin matrix formulations are also possible.

---

[92] Note we use U_2 in this example (for illustrative purposes) as opposed to the previous example where we used U_1.

7. Quantum Measurement Theory II

In chapter 2 we began our discussion of Quantum Measurement Theory covering the part of the theory that dealt with commuting eigenvalue operators. This part of Quantum Measurement Theory forms the basis of Classical Operator Logic, which is completely deterministic and has no aspect related to probability.

In this chapter we consider the part of Quantum Measurement Theory that deals with non-commuting sets of eigenvalue operators, and corresponding non-commuting sets of filter (measurement) projection operators. We will use the physical theory of this part of Quantum Measurement Theory in chapter 8 to develop Quantum Operator Logic.

7.1 Non-Commuting Eigenvalue Operators

In chapter 2 we denoted eigenvalue operators whose eigenvalues were possible physical values of a physical quantity as $A_i(a_{ij})$ where i labels the physical quantity and a_{ij} is its jth eigenvalue. A filter or measurement operator that filtered out all parts of a state except the part with eigenvalue a_{ij} at some stage of an experiment was denoted $M(A_i(a_{ij}))$. In this chapter we consider the case of sets of non-commutative eigenvalue operators. We assume that in general members of the set of eigenvalue operators do not commute:

$$[A_i(a_{ij}), A_k(a_{km})] \neq 0 \qquad (7.1)$$

and consequently, in general, members of the set of measurement operators do not commute:

$$[M(A_i(a_{ij})), M(A_k(a_{km}))] \neq 0 \qquad (7.2)$$

Eq. 7.2 implies that the order of the measurements at various stages of an experiment is significant. Change the order, and the experiment, and its results, change also.

7.2 Generalized Measurement Operators

The measurement operators that we considered in chapter 2 were filters for one eigenvalue: at some stage of an experiment $M(A_i(a_{ij}))$ filtered out all incoming states except those states with the eigenvalue a_{ij} for the A_i'th operator.

We now define a generalization of $M(A_i(a_{ij}))$ that, at some stage of an experiment, accepts states with A_i eigenvalue a_{ij} and outputs states whose A_i eigenvalue is a_{ik} for some value k – an eigenvalue transforming filter. We denote this operator

$$M(A_i(a_{ik}), A_i(a_{ij})) \qquad (7.3)$$

Note that

$$M(A_i(a_{ik}), A_i(a_{ij})) \neq M(A_i(a_{ij}), A_i(a_{ik})) \qquad (7.4)$$

in general since the order of the projections does matter when the set of eigenvalue operators is not commutative.

The multiplication rule for the $M(A_i(a_{ik})), A_i(a_{ij}))$ transformation filters is not commutative. If we consider two stages of these measurements, represented by

$$M(A_i(a_{in}), A_i(a_{im}))M(A_i(a_{ik}), A_i(a_{ij})) \qquad (7.5)$$

then nothing will reach the following stage of the experiment if $m \neq k$. However if $m = k$ then states (or parts of states) being passed to the projection for a_{im} will pass on to the next stage. Thus

$$M(A_i(a_{in}), A_i(a_{im}))M(A_i(a_{ik}), A_i(a_{ij})) = \delta(a_{im}, a_{ik})M(A_i(a_{in}), A_i(a_{ij})) \qquad (7.6)$$

Eq. 7.4 is a generalization of eq. (2.3.9). Note that by interchanging n and j in eq. 7.4 we see that multiplication of these M operators is not commutative in general.

7.3 Products of Generalized Measurement Operators

Eq. 7.6 gave the multiplication rule for eigenvalue transformation filters where the eigenvalues are those of the same operator. We now consider the case of transformation filters for differing eigenvalues and eigenvalue operators. Generalized measurement operators or transformation filters have the form:

$$M(A_i(a_{ik}), A_j(a_{jm})) \qquad (7.7)$$

This transformation filter operator, at some stage of an experiment, accepts an eigenstate of A_j with eigenvalue a_{jm} and then outputs an eigenstate of A_i with eigenvalue a_{ik} to the next stage. Note the definition implies

$$M(A_i(a_{ik}), A_i(a_{ik})) = M(A_i(a_{ik})) \qquad (7.8)$$

At successive stages of an experiment a product of transformation filters may occur[93]

$$M(A_i(a_{ik}), A_j(a_{jm}))M(A_a(a_{ab}), A_c(a_{cd})) \qquad (7.9)$$

Eq. 7.9 raises the issue of the "compatibility" of the eigenstate of A_a with eigenvalue a_{ab} that is output to the left transformation filter which only accepts (selects) an eigenstate of A_j with eigenvalue a_{jm}. The issue is resolved by noting 1) the right (first) transformation filter accepts eigenstates of A_c with eigenvalue a_{cd}; 2) the left (second) transformation filter outputs an eigenstate of A_i with eigenvalue a_{ik}; and 3) the second transformation filter will only accept, at best, a fraction of the output of the first transformation filter. Thus[94] one concludes

$$M(A_i(a_{ik}), A_j(a_{jm}))M(A_a(a_{ab}), A_c(a_{cd})) = <A_j(a_{jm})|A_a(a_{ab})>M(A_i(a_{ik}), A_c(a_{cd}))$$
$$(7.10)$$

---

[93] The filter products execute from right to left.
[94] Eq. 7.10 is an experimentally proven fact for quantum physics and thus is based on Nature, which cannot be mathematically inconsistent or incorrect.

where $<A_j(a_{jm})|A_a(a_{ab})>$ is assumed to be a number[95] that is a measure of the fraction of the states with the eigenvalue a_{ab} of operator A_a that can be accepted by the the filter for the state with the eigenvalue a_{jm} of operator A_j. The bra–ket notation $<... |... >$ used in eq. 7.10 was invented by Dirac. Eq. 7.10 is an assumption that has been verified repeatedly since Dirac's formulation of measurement theory in his 1931 book. Note that

$$<A_j(a_{jm})|A_j(a_{ab})> = \delta(a_{jm}, a_{ab}) \qquad (7.11)$$

which states the eigenvectors of a hermitean operator are defined to be orthonormal. $\delta(a_{jm}, a_{ab})$ is either a Kronecker δ-function (discrete eigenvalues) or a Dirac δ-function (continuous eigenvalue spectrum).

7.4 Probabilities and Generalized Measurement Operators

Eq. 7.10 introduces numbers, which will turn out to be the source of quantum probabilities. To see the origin of this interpretation we first need to relate transformation filters to measurement operators. The relationships emerge from eqs. 7.8 and 7.10:

$$M(A_i(a_{ik}))M(A_a(a_{ab}), A_c(a_{cd})) = M(A_i(a_{ik}), A_i(a_{ik}))M(A_a(a_{ab}), A_c(a_{cd}))$$
$$= <A_i(a_{ik})|A_a(a_{ab})>M(A_i(a_{ik}), A_c(a_{cd}))$$

$$\qquad (7.12)$$

$$M(A_i(a_{ik}), A_j(a_{jm}))M(A_a(a_{ab})) = <A_j(a_{jm})|A_a(a_{ab})>M(A_i(a_{ik}), A_a(a_{ab}))$$

Using eq. 2.3.13a eqs. 7.12 become

$$M(A_a(a_{ab}), A_c(a_{cd})) = \sum_k <A_i(a_{ik})|A_a(a_{ab})>M(A_i(a_{ik}), A_c(a_{cd}))$$

$$\qquad (7.13)$$

$$M(A_i(a_{ik}), A_j(a_{jm})) = \sum_b <A_j(a_{jm})|A_a(a_{ab})>M(A_i(a_{ik}), A_a(a_{ab}))$$

Eqs. 7.13 lead to a summation rule for the numbers $<A_j(a_{jm})|A_a(a_{ab})>$, namely,

$$\sum_b <A_j(a_{jm})|A_a(a_{ab})> <A_a(a_{ab})|A_k(a_{kn})> = <A_j(a_{jm})| A_k(a_{kn})> \qquad (7.14)$$

---

[95] Thus it commutes with all transformation filters.

Eq. 7.14 follows from expanding $M(A_j(a_{jm}))M(A_k(a_{kn}))$ using eqs. 7.8 and 7.12:

$$M(A_j(a_{jm}))M(A_k(a_{kn})) = <A_j(a_{jm})|A_k(a_{kn})> M(A_j(a_{jm}), A_k(a_{kn})) \qquad (7.15)$$

and using eqs. 2.3.13a and 7.12:

$$
\begin{aligned}
M(A_j(a_{jm}))M(A_k(a_{kn})) &= \sum_b M(A_j(a_{jm}))M(A_i(a_{ib}))M(A_k(a_{kn})) \\
&= \sum_b M(A_j(a_{jm}))M(A_i(a_{ib}), A_i(a_{ib}))M(A_k(a_{kn})) \\
&= \sum_b <A_j(a_{jm})|A_i(a_{ib})>M(A_j(a_{jm}),A_i(a_{ib}))M(A_k(a_{kn})) \\
&= \sum_b <A_j(a_{jm})|A_i(a_{ib})>M(A_j(a_{jm}),A_k(a_{kn}))<A_i(a_{ib})|A_k(a_{kn})> \\
&= \sum_b <A_j(a_{jm})|A_i(a_{ib})><A_i(a_{ib})|A_k(a_{kn})>M(A_j(a_{jm}),A_k(a_{kn}))
\end{aligned}
$$
$$(7.16)$$

Combining eqs. 7.15 and 7.16 gives

$$<A_j(a_{jm})|A_k(a_{kn})> = \sum_b <A_j(a_{jm})|A_i(a_{ib})><A_i(a_{ib})|A_k(a_{kn})> \qquad (7.17)$$

An important special case of eq. 7.17 is $j = k$. Then

$$\delta(a_{km}, a_{kn}) = \sum_b <A_k(a_{km})|A_i(a_{ib})><A_i(a_{ib})|A_k(a_{kn})> \qquad (7.18)$$

reflecting the orthonormality of the eigenvectors of A_k.

The "number" $<A_j(a_{jm})|A_a(a_{ab})>$ that first appears in eq. 7.10 appears to be the fraction of states $|A_a(a_{ab})>$ that transform to states $|A_j(a_{jm})>$. However there is a transformation under which the quantum measurement algebra is invariant that puts this interpretation of $<A_j(a_{jm})|A_a(a_{ab})>$ in doubt. For, if we let

$$<A_j(a_{jm})|A_a(a_{ab})> \rightarrow <A_j(a_{jm})|A_a(a_{ab})>\alpha(a_{ab})/\beta(a_{jm}) \qquad (7.19)$$

and

$$M(A_j(a_{jm}))M(A_a(a_{ab})) \rightarrow M(A_j(a_{jm}))M(A_a(a_{ab}))\beta(a_{jm})/\alpha(a_{ab}) \qquad (7.20)$$

for some α and β, throughout the prior development (eqs. 7.10 – 7.18), then we find the equations are unchanged (invariant). Thus the number $<A_j(a_{jm})|A_a(a_{ab})>$ is not, in itself, physically relevant since it can be scaled without affecting the basic equations. But the quantity

$$p(A_a(a_{ab}), A_j(a_{jm})) = <A_j(a_{jm})|A_a(a_{ab})><A_a(a_{ab})|A_j(a_{jm})> \qquad (7.21)$$

is invariant under the transformation eq. 7.19 and satisfies

$$p(A_a(a_{ab}), A_j(a_{jm})) = p(A_j(a_{jm}), A_a(a_{ab})) \qquad (7.22)$$

and

$$1 = \sum_b p(A_a(a_{ab}), A_j(a_{jm})) \qquad (7.23)$$

by eq. 7.18. The natural, and correct, interpretation of $p(A_a(a_{ab}), A_j(a_{jm}))$ is that it is the probability of a system originally in the state with eigenvalue a_{jm} of A_j to be filtered to the state with eigenvalue a_{ab} of operator A_a. Eq. 7.23 shows that the sum of the probabilities of all filtered output to eigenstates of A_a is unity as it should be. Since all eigenvalue operators are assumed to be hermitean

$$<A_j(a_{jm})|A_a(a_{ab})> = <A_a(a_{ab})|A_j(a_{jm})>^* \qquad (7.24)$$

where * indicates complex conjugate and thus eq. 7.21 can be expressed as the absolute value squared

$$p(A_a(a_{ab}), A_j(a_{jm})) = |<A_a(a_{ab})|A_j(a_{jm})>|^2 \qquad (7.25)$$

Eq. 7.25 implies all probabilities are positive and, taking account of eq. 7.23, we see that

$$0 \leq p(A_a(a_{ab}), A_j(a_{jm})) \leq 1 \qquad (7.26)$$

Thus the Quantum Measurement Theory inexorably leads to quantum probabilities. These results have been repeatedly verified over the past eighty years. They constitute the bedrock of physics and the foundation of Reality as we see it.[96]

---

[96] There are some who feel quantum theory is not correct or the result of a more fundamental theory. There is not one shred of reliable data to support that view.

7.5 Dirac's Bra–Ket Notation for Filters

Dirac's bra–ket notation for eigenstates and operators enables us to restate the results of previous sections in a simpler, more straightforward way.[97] Part of this notation has already been introduced in eq. 7.10 in the (generally complex) number $<A_j(a_{jm})|A_a(a_{ab})>$.

Now we proceed to define a ket-type eigenvector of operator A_a with eigenvalue a_{ab} as[98]

$$|A_a(a_{ab})> \tag{7.27}$$

and a bra-type eigenvector of operator A_a with eigenvalue a_{ab} as[99]

$$<A_a(a_{ab})| \tag{7.28}$$

The set of ket eigenvectors $|A_a(a_{ab})>$ for all eigenvalues a_{ab} span a Hilbert space and the set of bra eigenvectors $<A_a(a_{ab})|$ are the hermitean conjugates of the kets:

$$<A_a(a_{ab})| = (|A_a(a_{ab})>)^\dagger \tag{7.29}$$

An inner product of a bra and ket yields a complex number in quantum theory that is designated by $<A_j(a_{jm})|A_a(a_{ab})>$ and satisfies eq. 7.24. Note the redundant double $\|$ (that we will display for clarity below in some cases) is replaced by a single $|$ notationally.

We will now restate the results of the previous sections in this chapter in the bra-ket notation. Corresponding equations will be numbered as earlier in the chapter with "bk" appended to the equation number. For example, eq. 7.3 becomes eq. 7.3bk below.

First we begin with a generalization of $M(A_i(a_{ij}))$, an eigenvalue transforming filter, that, at some stage of an experiment, accepts states

---

[97] One could, in fact, say Dirac's bra–ket notation facilitates measurement operator manipulations just as Leibniz's differential notation facilitates differential calculus compared to Newton's "D" differential operator notation. A more complete discussion of Dirac's notation can be found in Dirac (1931), Gottfried (1989), and Messiah (1965).

[98] In terms of ordinary column vectors, row vectors and matrices $|A_a(a_{ab})>$ corresponds to a column vector.

[99] The state $<A_a(a_{ab})|$ corresponds to a row vector in terms of ordinary column vectors, row vectors, and matrices.

whose $A_i{}^{th}$ eigenvalue is a_{ij} and outputs states whose $A_i{}^{th}$ eigenvalue is a_{ik} for each value of k.

An eigenvalue transformation filter

$$M(A_i(a_{ik}), A_i(a_{ij})) \equiv |A_i(a_{ik})><A_i(a_{ij})| \qquad (7.3bk)$$

The expression $|A_i(a_{ik})><A_i(a_{ij})|$ can be viewed as a matrix generated by taking the outer product of a "column" vector with a "row" vector. Note that

$$|A_i(a_{ik})><A_i(a_{ij})| \neq |A_i(a_{ij})><A_i(a_{ik})| \qquad (7.4bk)$$

The multiplication rule for $M(A_i(a_{ik})), A_i(a_{ij}))$ is not commutative. If we consider two stages of measurements, represented by

$$|A_i(a_{in})><A_i(a_{im})||A_i(a_{ik})><A_i(a_{ij})| \equiv |A_i(a_{in})><A_i(a_{im})|A_i(a_{ik})><A_i(a_{ij})| \qquad (7.5bk)$$

then nothing will reach the next stage of the experiment if $m \neq k$. However if $m = k$ then the states being passed to the projection for a_{im} will be passed on to the next projection without hindrance. Thus

$$|A_i(a_{in})><A_i(a_{im})|A_i(a_{ik})><A_i(a_{ij})| = \delta(a_{im}, a_{ik}) |A_i(a_{in})><A_i(a_{ij})| \qquad (7.6bk)$$

since $<A_i(a_{im})|A_i(a_{im})> = 1$ by eq. 7.11.

Generalized measurement operator or transformation filter

$$M(A_i(a_{ik}), A_j(a_{jm})) \equiv |A_i(a_{ik})><A_j(a_{jm})| \qquad (7.7bk)$$

At successive stages of an experiment a product of transformation filters may occur. A product satisfies the equation:[100]

$$|A_i(a_{ik})><A_j(a_{jm})||A_a(a_{ab})><A_c(a_{cd})| = <A_j(a_{jm})|A_a(a_{ab})> |A_i(a_{ik})><A_c(a_{cd})| \qquad (7.10bk)$$

---

[100] This equation, and those following, show the efficacy of Dirac's notation in representing the measurement algebra.

The relationships implied by eqs. 2.3.13a, 7.8 and 7.10 are:

$$|A_i(a_{ik})><A_i(a_{ik})||A_a(a_{ab})><A_c(a_{cd})| = <A_i(a_{ik})|A_a(a_{ab})>|A_i(a_{ik})><A_c(a_{cd})|$$
$$(7.12bk)$$

$$|A_i(a_{ik})><A_j(a_{jm})||A_a(a_{ab})><A_a(a_{ab})| = <A_j(a_{jm})|A_a(a_{ab})>|A_i(a_{ik})><A_a(a_{ab})|$$

$$|A_a(a_{ab})><A_c(a_{cd})| = \sum_k <A_i(a_{ik})|A_a(a_{ab})>|A_i(a_{ik})><A_c(a_{cd})|$$
$$(7.13bk)$$

$$|A_i(a_{ik})><A_j(a_{jm})| = \sum_b <A_j(a_{jm})|A_a(a_{ab})>|A_i(a_{ik})><A_a(a_{ab})|$$

$$\sum_b <A_j(a_{jm})|A_a(a_{ab})> <A_a(a_{ab})|A_k(a_{kn})> = <A_j(a_{jm})| A_k(a_{kn})> \quad (7.14bk)$$

$$|A_j(a_{jm})><A_j(a_{jm})||A_k(a_{kn})><A_k(a_{kn})| = <A_j(a_{jm})|A_k(a_{kn})>|A_j(a_{jm})><A_k(a_{kn})|$$
$$(7.15bk)$$

$$|A_j(a_{jm})><A_j(a_{jm})||A_k(a_{kn})><A_k(a_{kn})| =$$
$$= \sum_b <A_j(a_{jm})|A_i(a_{ib})><A_i(a_{ib})|A_k(a_{kn})>|A_j(a_{jm})><A_k(a_{kn})|$$
$$(7.16bk)$$

$$<A_j(a_{jm})|A_k(a_{kn})> = \sum_b <A_j(a_{jm})|A_i(a_{ib})><A_i(a_{ib})|A_k(a_{kn})> \quad (7.17bk)$$

The above equations in Dirac's notation have the same visual superiority over the M notation that Leibniz's differential notation had over Newton's D derivative notation. Therefore we will use Dirac's notation in the remainder of this book.

8. Quantum Operator Logic

8.1 Quantum Operator Logic is Not Fuzzy Logic

Fuzzy Logic in its simplest form is a type of <u>classical</u> probabilistic logic that is based on assigning probabilities to the members of the domain of a predicate that leads to a probability for each statement consisting of the predicate and one or more of the subjects of its domain.

The Quantum Operator Logic that we will describe in this chapter is Quantum in nature and thus is not related to Fuzzy Logic.

8.2 Predicates are Filters

In the preceding chapter we considered the formalism for Quantum Measurement Theory based on the correspondence between the stages of a quantum experiment and the sequence of parts of a statement. In part this analogy is based on Hilbert's[101] view of a predicate, "the use of the term 'predicate' is the one usual in philosophy, namely, that by which one subject can be more particularly characterized"; and the analysis of a quantum experiment as a series of filtrations which particularize the input to an experiment through its sequence of stages to reach a final stage of output.

8.3 Quantum Operator Logic Formalism

Quantum Operator Logic formalism is an extension of the formalism developed for classical Operator Logic in chapters 3 and 4 to the case where the eigenvalue operators of the terms in a universe commute with each other[102] but *the set of truth value projection operators of the universe of discourse do not commute with each other in general.*

---

[101] Hilbert (1950) p. 44.

[102] This enables the eigenvalue operators all terms in a universe of discourse to have a simultaneous eigenvector, namely, the universe state as defined previously.

We will define the state of a quantum universe of discourse in the same manner as we did in the case of the classical universe of discourse using Dirac's bra-ket notation for states and operators. We will assume we have q primitive terms with which any statement in the universe can be expressed.

Thus the universe state will be

$$|\Omega_u> = \prod_{i \varepsilon \{m\}} |E_i> \tag{8.1}$$

$$= |E_1, E_2, E_3, \dots, E_q> \tag{8.2}$$

where $\{m\} = \{1, 2, 3, \dots, q\}$ and

$$|E_1, E_2, E_3, \dots, E_q> = \prod_{i=1}^{q} |E_i> \tag{8.3}$$

for the set $\{m\}$ consisting of q non-zero strings/symbols, E_i, with values of $gn(E_i)$. Note $|\Omega_u>$ is normalized to one: $<\Omega_u|\Omega_u> = 1$.

In defining eqs. 8.1 and 8.3 we are assuming that the set of q eigenvalue operators for the primitive terms, which we denote $\{A_i\}$, is compatible—the eigenvalue operators all commute with each other. On the other hand elements of the set of projection operators has the form[103]

$$P_{E_k} = P'_{E_k} \tfrac{1}{2}[1 + c_{E_k}{}^\dagger c_{E_k}] \tag{8.4}$$

where P'_{E_k} is a truth projection operator. The set of truth projection operators P'_{E_k} form an incompatible set—some, or all, of them do not commute with each other in general. The set of operators with the form $\tfrac{1}{2}[1 + c_{E_k}{}^\dagger c_{E_k}]$, which also appear in eq. 8.4, do commute with each other and form a compatible set.

---

[103] These operators, together with the eigenvalue operators, conform to the general character of quantum mechanics Hilbert space operators. Truth projection operators are well-defined Hilbert space operators and thus lie within the framework of quantum theory.

The operators $c_{E_k}^\dagger$ and c_{E_k} are raising and lowering operators defined by eq. 6.5. (We will not use the spinor formalism here to avoid mathematical complexity – although we could have used it.)

The probability of the truth of a statement:

$$\text{statement} = \langle \Omega_u | A_1 + A_2 + A_3 + \ldots | \Omega_u \rangle \qquad (8.5)$$

is determined by the evaluation of its probability amplitude

$$\text{Probability Amplitude } = \langle \Omega_u | P_{1tot}\, P_{2tot}\, P_{3tot} \ldots | \Omega_u \rangle \qquad (8.6)$$

where the $P_{itot} = P'_i P_i$ is the projection operator corresponding to the eigenvalue operator A_i. For a predicate term A_{E_k} the truth projection operator factor P'_{E_k} has the form

$$P'_{E_k} = \sum_j a_{E_j E_k} |E_j\rangle\langle E_k| \qquad (8.7)$$

where the numbers $a_{E_j E_k}$ are quantum probability amplitudes. Each combination of subject, E_j, and predicate, E_k, in a statement or clause has the quantum probability $|a_{E_j E_k}|^2$ of being true.

For a subject term A_{E_j} the truth projection operator factor is

$$P'_{E_j} = |E_j\rangle\langle E_j| \qquad (8.8)$$

A clause (or simple subject-predicate statement) has the form

$$\begin{aligned}
\text{statement} &= \langle \Omega_u | A_{\text{subject}} + A_{\text{predicate}} | \Omega_u \rangle \qquad (8.9) \\
&= \text{gn(subject)gn(predicate)}
\end{aligned}$$

namely, a sequence of two Gödel numbers equivalent to two terms in the universe of discourse. The probability amplitude is (from eqs. 8.7 and 8.8)

$$\begin{aligned}
\text{Probability Amplitude} &= \langle \Omega_u | P_{\text{subject}} P_{\text{predicate}} | \Omega_u \rangle \qquad (8.10) \\
&= \langle \Omega_u | P'_{\text{subject}} P'_{\text{predicate}} | \Omega_u \rangle \\
&= \langle \Omega_u | |E_{\text{subject}}\rangle\langle E_{\text{subject}}| a_{\text{subject,predicate}} |E_{\text{subject}}\rangle\langle E_{\text{predicate}}| \,|\Omega_u\rangle \\
&= a_{\text{subject,predicate}}
\end{aligned}$$

where $P'_{subject}$ and $P'_{predicate}$ are the eigenvalue operators' truth value projections. The probability of the statement being true is

$$p_{subject,predicate} = |a_{subject,predicate}|^2 \qquad (8.11)$$

Quantum probability amplitudes are assigned to all subjects in the domain of a predicate according to some quantum rule or calculational scheme.

Note on Universes of Discourse with Sets of Incompatible Quantum Operator Terms

It is possible for a universe of discourse to exist with a set of incompatible term eigenvalue operators.[104] A classic example of such a case would be the statement

$$x + p = p + x$$

where x is the position operator of a particle and p is its momentum. The Heisenberg Uncertainty Principle states that x and p are incompatible operators.

A universe of discourse with a set of incompatible eigenvalue operators is handled in essentially the same way as discussed above. The most important restriction is that the order of the terms in a statement and the order of the corresponding eigenvalue operator expression is significant and cannot be changed without changing the meaning of the statement in general.

The procedure is to map each term in a statement to a Gödel number as before. *Just as we mapped c-number terms to Gödel numbers we can map q-number terms to Gödel numbers*[105]. Then to transform the statement into an eigenvalue operator expression and evaluate its universe state[106] expectation value.

$$<\Omega_u|A_1 + A_2 + \ldots|\Omega_u>$$

---

[104] The source of this incompatibility could be that the universe of discourse is inherently quantum such as a quantum mechanical harmonic oscillator system or Quantum Electrodynamics. In cases of this sort the statements consist of operators not ordinary c-number terms. Therefore the operator terms of the statement do not necessarily commute ab initio.

[105] Individuals uncomfortable with mapping q-number terms to c-numbers (Gödel numbers) should remember that the path integral formulation of quantum field theories uses c-number fields and obtains the same results as perturbative expansions of the corresponding q-number quantum field theory. A study of the path integral formulation of quantum field theory from the perspective of a mapping to Gödel numbers would appear to be an interesting endeavor.

[106] Eq. 3.4.3.

The truth value probability amplitude of a statement is given by

$$<\Omega_{uq}|F_q(P'_1, P'_2, ...)(P_1 P_2 ...)|\Omega_{uq}>$$

where $F_q(P'_1, P'_2, ...)$ is the truth value expression that generates the quantum probability amplitude and the state $|\Omega_{uq}>$ is

$$|\Omega_{uq}> = |\Omega_u>|\Omega_q>$$

where $|\Omega_q>$ is a quantum state describing a physical (or other) state and $|\Omega_u>$ is given above.

This form generates a specific truth value probability amplitude for the statement for the specific quantum state $|\Omega_q>$. An example of this type of quantum expression is given in section 8.6 below.

This form is of particular interest because it opens the possibility of evaluating the probability of a statement being true in the wider context of a set of possible forms of the statement. For example, a statement might have a parameter in it that can take a range of possible values. If the knowledge base of the statement specifies a quantum probability function F_q for the value of the parameter then we can obtain a truth value probability amplitude for specific values of the parameter. Thus we can see the range of probabilities for various values of a parameter in a physical law. An example based on Quantum Electrodynamics is given in section 8.6.

Returning to the case where all eigenvalue operators commute we see we have a sentential calculus and a rule to reduce all compound statements to equivalent series of clauses as in chapter 3.

We note again the sentential connectives and their interpretation

Connective	Symbol
and	&
or	v
if ... then (modus ponens)	→
if and only if	~
not	___

Denoting sentences with upper case letters: A, B, ... the five basic sentential combinations have the quantum probabilities:

1. A & B is a compound sentence whose probability is the product of the probabilities of A and B.
2. A v B is a compound sentence whose probability is the sum of the probabilities of A and B.
3. A → B is a compound sentence whose probability = 1 − (probability of A)·(1 − (probability of B)) based on eq. 3.8.14. This is based on the non-probabilistic case: A → B is false if and only if A is true and B is false, which is equivalent to $\overline{AB}$ or in words is equivalent to NOT(A and NOT B).
4. A ~ B is a compound sentence whose probability = 1 − |(probability of A) − (probability of B)|. This is based on the non-probabilistic case: A ~ B is true if and only if both A and B are true, or both A and B are false.
5. The sentence $\underline{A}$ has the probability = 1 − (probability of A). This is based on the non-probabilistic case: the sentence $\underline{A}$ is false if A is true, and is true if A is false.

$$(8.12)$$

In specifying these probability rules we assume that if a statement has the probability p of being true then $p_f = 1 - p$ is the probability that the statement is false.

8.3.1 Probabilistic Example "if … then …" Modus Ponens

We will now examine the case of a probabilistic modus ponens (if … then …) statement. Suppose we have a statement

$$A \rightarrow B \quad \text{equivalent to} \quad \underline{AB} \qquad (8.13)$$

and A can be expressed as a sequence of anded clauses:

$$A = \text{clauseA1 and clauseA2 and …} \qquad (8.14)$$

and B can also be expressed as a sequence of anded clauses:

$$B = \text{clauseB1 and clauseB2 and …} \qquad (8.15)$$

Then the probability of the statement eq. 8.13 according to eq. 8.12 is given by the expression:

probability = 1 − (probability of A)·(1 − (probability of B))   (8.16)

where the probability amplitude for A is[107]

$<\Omega_u|A|\Omega_u> = |<\Omega_u|clauseA1|\Omega_u><\Omega_u|clauseA2|\Omega_u>...|\{1 −$
$− 2\theta(|<\Omega_u|clauseA1|\Omega_u>| − <\Omega_u|clauseA1|\Omega_u> + |<\Omega_u|clauseA2|\Omega_u>| −$
$− <\Omega_u|clauseA2|\Omega_u> + ...)\}$ (8.17)

by eq. 3.8.10 and where the probability amplitude for B is

$<\Omega_u|B|\Omega_u> = |<\Omega_u|clauseB1|\Omega_u><\Omega_u|clauseB2|\Omega_u>...|\{1 −$
$− 2\theta(|<\Omega_u|clauseB1|\Omega_u>| − <\Omega_u|clauseB1|\Omega_u> + |<\Omega_u|clauseB2|\Omega_u>| −$
$− <\Omega_u|clauseB2|\Omega_u> + ...)\}$ (8.18)

The probability of A is

$$|<\Omega_u|A|\Omega_u>|^2$$ (8.19)

and the probability of B is

$$|<\Omega_u|B|\Omega_u>|^2$$ (8.20)

If, for the sake of example,

$$|<\Omega_u|A|\Omega_u>|^2 = \tfrac{3}{4}$$ (8.21)

and

$$|<\Omega_u|B|\Omega_u>|^2 = \tfrac{1}{4}$$ (8.22)

then the probability of A → B

$$p(A \to B) = 1 − \tfrac{3}{4}(1 − \tfrac{1}{4}) = 7/16$$ (8.23)

For the special case where the probability of A equals one and the probability of B is one then p(A → B) = 1. If the probability of A equals one and if the probability of B is zero then p(A → B) = 0. Both these special cases conform to our expectations based on classical Logic.

---

[107] Expressions of the form $<\alpha|O|\beta>$ are inner products that are called *expectation values* of the operator O in quantum mechanics and are often denoted as $(\alpha, O\beta)$ in mathematics texts. In this example and the following example we factor expectation values into products of expectation values of clauses since this procedure is both well defined and produces results consistent with our intuitive expectations. A somewhat similar procedure is followed in perturbation theory expansions in quantum field theory. In perturbation theory, terms with more than two operators in them are expanded in terms of vacuum expectation values of products of two quantum field operators.

8.4 The Probability Amplitudes of Subjects in Predicate Domains

Probabilistic Operator Logic statements range from true (probability = 1); to infinitesimal probabilities p of being true with $0 < p \leq 1$. In the definition of a predicate projection operator (eq. 8.7) only non-zero terms appear—one term for each subject in the subject domain of the predicate. All probabilities associated with subjects in the domain of a predicate are therefore non-zero although they can be arbitrarily small.

8.4.1 Probability of an Undecidable or a Nonsense Statement

Subjects outside the domain of a predicate must have probability zero since probabilities are necessarily non-negative. Since nonsense statements and undecidable statements are outside the domain of a predicate they have probability zero. The domain of a predicate in Quantum Operator Logic is therefore an open set whose members in combination with the predicate have probabilities p satisfying $0 < p \leq 1$.

These conclusions are consistent with our view of experiments within the framework of Quantum Measurement Theory. The physical equivalent of an undecidable statement (or a nonsense statement) is an experiment that has a series of stages that lead to a certain result with probability zero.[108]

If we reconsider the case of Classical Operator Logic as presented in chapter 3 we see that the product of a subject projection operator from the subject domain of a predicate with the projection operator of the predicate was non-zero, while the product of a subject projection operator for a subject outside the domain of a predicate with the projection operator of the predicate was zero.

8.4.2 Quantum Probabilistic Logic Example

Consider a complex quantum mechanical process that causes a stream of photons to fill a circle of a certain radius on a screen. The probability amplitudes of the various possible radii of the circle are

---

[108] In the quantum sector undecidable statements have probability zero. By grouping them with false statements we implement the rule that undecidable statements are assigned the value false. This rule is unambiguous and well defined. It is also consistent with Blaha's proof that Gödel's Theorem implies Nature must be quantum. (Blaha 2005c)

calculated and found to yield the $a_{subject,predicate}$ values for the cases listed below.

Predicate = "is the radius"

Domain:

Subject1 = "Five cm" $\qquad a_{subject,predicate} = \frac{1}{2}$

Subject2 = "Ten cm" $\qquad a_{subject,predicate} = \frac{1}{2}$

Subject3 = "Fifteen cm" $\qquad a_{subject,predicate} = \frac{1}{2}$

Subject4 = "Twenty cm" $\qquad a_{subject,predicate} = \frac{1}{2}$

Based on this universe of discourse we find the probabilities of each statement to be:

p("Five cm ", "is the radius") = ¼
p("Ten cm ", "is the radius") = ¼
p("Fifteen cm ", "is the radius") = ¼
p("Twenty cm ", "is the radius") = ¼

The probabilities for each case happen to be equal and sum to one as they should.

8.5 Quantum Subuniverses Generated by Lowering Operators

In the preceding sections we have considered Quantum Probabilistic Universes based on Quantum Operator Logic, which is in turn based on Quantum Measurement Theory as developed in chapters 3 and 7. We now consider the possibility of Quantum Subuniverses with the ultimate view of considering quantum mathematical deductive theories of the fundamental laws of the universe.

The creation of subuniverses of discourse is described in the case of classical Operator Logic in sections 5.2 and 5.3. The creation of a quantum subuniverse follows the same procedure of deleting primitive terms from a universe of discourse by using lowering operators as in those sections.

The quantum subuniverse so created then is treated as in prior sections of this chapter to obtain the probabilities of statements.

8.6 Quantum Electrodynamics as a Universe with Incompatible Terms

In section 5.4 we considered a generalization of the classical Maxwell equations in the absence of sources and currents:

$$\nabla \times \mathbf{E} = - \beta \partial \mathbf{B}/\partial t$$
$$\nabla \cdot \mathbf{E} = 0 \qquad\qquad (5.4)$$
$$\nabla \cdot \mathbf{B} = 0$$
$$\nabla \times \mathbf{B} = - \beta \partial \mathbf{E}/\partial t$$

where β is a constant and $\times$ represents the vector cross product while $\cdot$ represents a vector inner product. If $\beta = 1$, then eqs. 5.4 are exactly the Maxwell equations of electromagnetism in the absence of sources and currents.

We now consider the quantum field theory corresponding to eqs. 5.4, which is a generalization of Quantum Electrodynamics (QED) in the absence of sources and currents.

As before we have nine primitive terms to consider:

"$\nabla \times$"   "$\nabla \cdot$"   "$=$"   "$-$"   "β"   "$\partial/\partial t$"   **"E"**   **"B"**   "0"

which we associate with the eigenvalue operators A_1, A_2, ... , A_9. However the field operators **E** and **B** do not commute. So we define the quantum universe state $|\Omega_u\rangle$ as

$$|\Omega_u\rangle = |\text{ "}\nabla \times\text{", "}\nabla\cdot\text{", "}=\text{", "}-\text{", "}\beta\text{", "}\partial/\partial t\text{", "E", "B", "}0\text{"}\rangle$$
$$(8.24)$$

with string (Gödel number) eigenvalues. Then the first classical pseudo-Maxwell equation

$$\nabla \times \mathbf{E} = - \beta \partial \mathbf{B}/\partial t$$

is represented by

$$\langle\Omega_u|(A_1 + A_7 + A_3 + A_9)|\Omega_u\rangle \equiv \nabla \times \mathbf{E} = - \beta \partial \mathbf{B}/\partial t$$
$$(8.26)$$

The truth value probability amplitude of the statement requires us to specify a projection function $F_q(P'_1, P'_2, ...)$ whose form is follows from the knowledge base of the universe, and to specify a quantum state that also follows from the knowledge base of the universe.

For illustrative purposes only, we will select the projection function $F_q(P'_1, P'_2, ...) = 1$ and the quantum state to be the ground state of a one-dimensional simple harmonic oscillator where β is a "spatial" dimension and the "potential energy" is $V = \frac{1}{2}k(\beta - 1)^2$. The ground state wave function is

$$\Psi(\beta) = \pi^{-\frac{1}{4}} \exp[-\frac{1}{2}(km)^{\frac{1}{2}}(\beta - 1)^2] \qquad (8.27)$$

Thus the truth value probability amplitude as a function of β is

$$<\Omega_q| F_q(P'_1, P'_2, ...)|\Omega_q> \rightarrow \Psi^*(\beta)1\Psi(\beta) = \pi^{-\frac{1}{2}}\exp[-(km)^{\frac{1}{2}}(\beta - 1)^2]$$

and the probability that the statement is true for the value β is

$$p(\beta) = \pi^{-1}\exp[-2(km)^{\frac{1}{2}}(\beta - 1)^2] \qquad (8.28)$$

For large k the probability distribution is sharply peaked at $\beta = 1$, the value in the Maxwell equations. If this illustrative example were true in reality then experimental tests of the equations of electromagnetism would yield the Maxwell equations as the most probable result with other values of β appearing in experiments with much lower probability. It is important to note that $\beta = 1$ exactly in all known electromagnetic experiments and so this example is for illustrative purposes only.

8.6.1 Probability Distributions for Theories

The illustrative example provided above embodies an approach to determining numerical constants that frequently appear in physical theories. All such numerical constants are known to a certain degree of accuracy. Some are determined by symmetry conditions. But there is some uncertainty associated with every physical constant. Even such an accurately determined constant as the electromagnetic fine structure constant there is uncertainty in its value and, from time to time, speculations that its value is time dependent.

The approach to the calculation of the probability that β had a certain value assumed the knowledge base of the universe specified a means to determine the probability – a "pre-theory", if you will, that yields the probability distribution of values of β.

Such an approach might be of value in determining the many constants appearing in the Standard Model of Particles although the details of such an approach are lacking at present.

More generally, the question of the dimensions, symmetries, and terms in the dynamical equations of a theory might be addressed by a comprehensive pre-theory. In the case of quantum field theories the issues are muddied somewhat by changes in constants that take place due to renormalization effects, and mechanisms such as the Higgs Mechanism that change the values of masses. Thus it would appear that a successful approach along these lines is not likely in the near future. The axiomatic derivation of the Standard Model in Blaha (2008) appears to be the most promising current approach in the view of this author.

9. From Operator Logic to Reality - The Standard Model

9.1 Platonic Conception of the Relation of Ideas to Reality

Plato and subsequent Platonic philosophers postulated that there existed a realm of Ideas.[109] Each thing, both material and conceptual, in Reality had an abstract counterpart in the realm of Ideas that embodied its features. Thus there was an Idea of a plow, an Idea of Justice, and so on. The realm of Ideas was connected through mathematics (Number) to the realm of Reality with which we are familiar.

Plato and subsequent philosophers were not familiar with the modern view of Reality, which has only became apparent in the quantum revolution of the twentieth century. The essence of quantum theory has been distilled into Quantum Measurement Theory. This theory applies to non-relativistic quantum mechanics and relativistic quantum field theory, as well as SuperString theories.

Quantum Measurement Theory is a formal method of viewing the world. It describes the process of experimental measurement as a series of filtrations that occur at various stages of an experiment. Since all observations are experiments, it applies to all the ways that we obtain information about the universe in which we live. In the macroscopic world quantum effects are usually negligible. In the very small, quantum effects dominate physical processes. But the theory of Quantum Measurement applies in principle in all cases.

So the gropings of Plato and other philosophers towards a total view of Reality had an element of truth in that there is a fundamental set of Ideas that govern Reality, but was wrong—for good reason—Reality does not map directly to Ideas in a one-to-one fashion.

Instead, the set of Ideas is the set of laws of Quantum Measurement Theory (possibly supplemented by yet unknown laws). And the phenomena of Reality behave according to the laws of Quantum Measurement Theory whenever observations (experiments) are made.

---

[109] Plato discusses the concept of Ideas in the dialogue Parmedides as well as other dialogues.

The mathematics connecting Ideas to Reality are the probabilities emerging from the application of Quantum Measurement Theory to experimental situations. Thus the form of Platonic thought was correct in a general sense; but the specifics were incorrect due to the primitive state of physical science in those times.

9.2 Operator Logic Exists Independent of Our Knowledge of It

We constructed classical and quantum Operator Logic in the "language" of Hilbert space using the concepts and laws of Quantum Measurement Theory. Thus Operator Logic is based on concepts and laws that exist independently of our knowledge of them.[110] After all, quantum physics only emerged in the twentieth century yet the universe followed the laws of quantum theory since its beginning. Thus we must attribute a reality to the concepts and laws of Quantum Measurement Theory. Our philosophical stance must be Platonist—the concepts and laws of Quantum Measurement Theory have a true existence outside of our knowledge of them.[111]

The development of Operator Logic based on the concepts and laws of Quantum Measurement Theory gives Operator Logic the same status—a true existence outside of our knowledge of them.

9.3 Being and Existence of the Material World

The form of existence of Quantum Measurement Theory and Operator Logic is clearly that of the realm of Ideas. On the other hand space and time and matter and radiation exist as parts of Reality. Although Reality conforms to the laws of Quantum Measurement Theory, and Operator Logic, it has something extra: being (existence). The components of Reality are tangible—perceivable by our senses (directly or indirectly), and the components of Reality interact with each other. But how do the components of Reality acquire existence?

---

[110] The deepest part of our current physical understanding of the universe is Quantum Measurement Theory. All branches of physics ultimately conform to it. It is extremely unlikely that Quantum Measurement Theory will change.

[111] This statement should not be confused with the quantum mechanical requirement of an observer to make quantum mechanical measurements. Quantum physical processes can proceed without an observer although results, being probabilistic, can only be determined by an observer. We discuss observers in Operator Logic and Reality in appendix A.

Remarkably, there is strong evidence that the universe emerged from a point (or a "small" neighborhood of a point). So the question of being may degenerate to the question of "being" at a single point without extension, without space or time, and perhaps without content. A number of theorists have suggested that the Big Bang is a quantum fluctuation—something emerging from nothing (the "vacuum") in such a way that the sum total of the emergent fluctuation has zero energy and thus is still nothing if considered in toto.[112]

In view of the major uncertainties, both physically and philosophically, in the understanding of existence and being the question of the origin of being is primarily work for the future. Appendix C provides a preliminary discussion of being based on our knowledge of particle physics – a subject which confronts existence and non-existence (being and non-being) on an everyday experimental and theoretical basis.

9.4 The Origin and Mechanism of Physical Laws

When we were developing Operator Logic in earlier chapters we assumed a corpus of predicates and subjects, and also domains for each predicate. The specification of predicate domains was based on an implicit *knowledge base* that was used to determine the subjects that were in the domain of a given predicate and/or to determine the probability amplitude that a subject was in the domain of a particular predicate. Thus a knowledge base is an essential ingredient of Operator Logic.

In the case of physical Reality a physics knowledge base was implicit that operated between the various stages of an experiment and affected the filtrations that took place as the experiment progressed. Physical quantities such as mass, energy, momentum and so on, and the laws that governed them, constitute a knowledge base that is implicit in the application of Quantum Measurement Theory to physics.

To make the transition from the realm of Ideas to Reality not only do we need the existence of physical things but we also need a source for physical laws and a mechanism for their execution. In human societies legislatures frame laws and the police enforce them. In Reality we must determine the origin of physical laws and the mechanism by which they are "enforced" on the components of Reality.

---

[112] Being then becomes an illusory artifact of our consciousness.

The typical methods, by which physical laws are determined, are 1) a fundamental lagrangian which through "canonical" methods leads to the equations embodying physical laws; or 2) a set of fundamental equations from which all physical laws are derived.

In either of these cases the question arises: "Where did the fundamental lagrangian or equations come from?" One could say that their origin is not knowable within a scientific framework. (Perhaps a deity specified them.) Or one could say they were the result of some as yet unknown process of self-organization (order emerging from chaos).

Once we have the laws then the question arises of how things in Reality are coerced into obeying them. Who are the police that enforce physical laws? The importance of this question is very apparent today in the case of quantum entanglement, and especially in the Einstein-Rosen-Podolsky thought experiment. In this experiment doing something to a particle that is quantum entangled with another particle at a great distance away—affects the distant particle. How? Clearly there is an enforcement issue for physical laws that is not as yet understood. Interestingly, the enforcement must be instantaneous and thus infinitely faster than the speed of light. One mechanism that could resolve this problem is the tachyon propagation of quantum effects.[113] Tachyons can travel at infinite speed and thus instantaneously implement physical laws.

Thus we add the need for a source of physical laws and a mechanism for their enforcement to the need for the origin of Reality.

9.5 Connection to the Standard Model of Particles

The four aspects of Reality, space, time, matter, and radiation, are intimately related. Without matter or radiation then space and time are without meaning. Without change, time becomes unmeasurable and irrelevant. Without "things"[114] distance and thus space becomes unmeasurable and also irrelevant. So the presence of matter and/or radiation are necessary to make time and space meaningful and thus real.

They also modify the properties of time and space as Einstein's Theory of General Relativity demonstrates. Thus we are aware of the interrelations of these elements of Reality.

---

[113] See Blaha (2008) for a discussion of tachyon quantum field theory and the possible role they play in the Standard Model of elementary particles.
[114] Matter or radiation.

9.5.1 Matter

Our understanding of matter has evolved tremendously in the past two millenia. For much of that time we viewed matter as concrete, substantial "stuff" that we could kick as Dr. Johnson did to refute Bishop Berkeley's claim that matter was not "real". Dr. Johnson reputedly said, " I refute it thus!" as he kicked a rock. Starting in the nineteenth century it became clear that matter was composed of molecules, which in turn were composed of atoms, which in turn were composed of electrons circling a nucleus containing protons and neutrons (the solar system view of the atom.) In the nineteen seventies it became clear that protons, neutrons and other particles were composed of quark particles. So matter appears, at present, to be composed of eighteen types of quarks and six particles[115] called leptons. The major distinguishing property between quarks and leptons is that quarks experience a type of force called the strong interaction while leptons do not. Rather than describe quarks and leptons in detail we suggest the interested reader read one of the many popular books on elementary particles.

This author has proposed a set of postulates from which the theory of these particles (the Standard Model of Elementary Particles) may be derived.[116]

9.5.2 Matter is Insubstantial

We now wish to get to the heart of the nature of matter. What is a particle made of? It appears that a particle is made of form without substance (as we commonly understand substance). This form is particulate in part and wave-like in part. We might think of it as "nothing" upon which a semi-permanent form is imposed with the quality of "being." Being consists of existence for some interval of time and of some form of observability. An entirely unobservable being could have a persistent form but lacking observability could not be considered real since there would be no manifestations of it in our Reality. Fortunately, for us Quantum Theory requires an observer for all real things and events. Thus all known particles interact and that makes them part of our Reality.[117]

---

[115] Electrons, muons, τ particles – each with a corresponding type of neutrinos.

[116] Blaha (2008). A technical book with some "popular" chapters.

[117] This includes dark matter which interacts with normal matter through gravitation.

Particles exist and possess being. In addition we have found that particles can interact with each other to truly create new particles or to truly annihilate into radiation. So the forms imposed on nothing (particles) can transform to other forms but do so in such a way that certain features of these forms are preserved. These features satisfy what we call conservation laws such as the conservation of energy.

It is rather remarkable that particles can undergo true creation and annihilation because these features represent transformations between *being* and *not being*. Earlier we developed Operator Logic based on Quantum Measurement (Transformation) Theory and, after showing that one thus has a framework applicable to both Logic and Reality, we raised the issue of the "addition" of the quality of being (existence) to "make" Reality from our abstract formalism. Seeing the process of true creation and annihilation in the laboratory it is clear that being is an acquirable property. As such, the Big Bang, which seems to be the origin of the universe, can be accepted as a transition from nothing into being through an expansion from a point to space-time coupled to a fluctuation that led to a differentiation into regions of differing forms (types) of particles. Thus we perceive the nature of the process although the precise details remain to be determined.[118]

9.5.3 The Knowledge Base of Reality

As we have seen, a calculus universe of discourse consists of a set of axioms written in terms of undefined primitive terms and the theorems derived therefrom. A semantic universe of discourse attributes meanings/interpretations to the primitive terms thereby giving meaning to the axioms and consequent theorems. The Operator Logic/Quantum Measurement Theory formalism must be supplemented by a set of axioms for either type of universe of discourse, and also a physical interpretation of the primitive terms for a semantic universe of discourse. We will call the set of primitive term definitions and the set of axioms the *knowledge base* of the universe of discourse of Reality.

The determination of the knowledge base of physical Reality is beyond the scope of this book but one possibility is that expounded in Blaha (2008). Blaha (2008) lists a set of axioms that lead directly, and *exactly*, to the established features of the Standard Model of Elementary

---

[118] Blaha (2004) describes a theory of a quantum Big Bang.

Particles. The features of this theory, that were determined over a period of forty-five years (1930 – 1975 approximately) of experimental and theoretical work. The features include parity violation, peculiar symmetries, and the complex nature of the particle spectrum. These features are *exact* results of the axioms in Blaha (2008) unlike all other theoretic attempts to explain the Standard Model. Those attempts all view the Standard Model as a low energy approximation of a larger theory and have no inherent justification for parity violation but rather frame their theories to incorporate it.

However we will not commit to a specific knowledge base for Reality in this book but will simply proceed to relate our Operator Logic/Quantum Measurement Theory formalism to the basic ingredients that any knowledge base must have.

We know that every basic particle of matter, whether quark or lepton, has spin ½. We have seen in chapter 6 (and particularly in section 6.3) that the basic algebra of eigenvalue operators, and that of the raising and lowering operators, is the same as the algebra of spin ½ particles. Thus we have a rationale for requiring the knowledge base to have spin ½ particles.

The non-physicist reader may wish to review section 6.3 at this point, as we develop an approach to the spin ½ particles of Reality based on the spinor formulation of quantum Operator Logic. We will start with the direct sum of two spinor universes of discourse as given in eqs. 6.13 – 6.18. In particular the 4-vectors of four rows were

$$S_{\uparrow\uparrow k} = \begin{bmatrix} S_{1k\uparrow} \\ \\ S_{2k\uparrow} \end{bmatrix} \quad S_{\uparrow\downarrow k} = \begin{bmatrix} S_{1k\uparrow} \\ \\ S_{2k\downarrow} \end{bmatrix} \quad S_{\downarrow\uparrow k} = \begin{bmatrix} S_{1k\downarrow} \\ \\ S_{2k\uparrow} \end{bmatrix} \quad S_{\downarrow\downarrow k} = \begin{bmatrix} S_{1k\downarrow} \\ \\ S_{2k\downarrow} \end{bmatrix} \qquad (6.15)$$

We form linear combinations of these vectors to establish contact with the standard treatment of spin ½ free quntum fields in physics:

$$u(k, +\tfrac{1}{2}) = \begin{bmatrix} 1 \\ 0 \\ 0 \\ 0 \end{bmatrix} \quad u(k, -\tfrac{1}{2}) = \begin{bmatrix} 0 \\ 1 \\ 0 \\ 0 \end{bmatrix} \tag{9.1}$$

and

$$v(k, -\tfrac{1}{2}) = \begin{bmatrix} 0 \\ 0 \\ 1 \\ 0 \end{bmatrix} \quad v(k, +\tfrac{1}{2}) = \begin{bmatrix} 0 \\ 0 \\ 0 \\ 1 \end{bmatrix} \tag{9.2}$$

where the first arguments of u and v, which have the value $k = 0$, will be identified as the momentum of a spin ½ particle shortly. The second arguments of u and v are the values of particle spin: +½ represents an "up" spin state and −½ represents a "down" spin state.[119] Note the relations to the spinors in eq. 6.15 such as

$$u(k, +\tfrac{1}{2}) + v(k, -\tfrac{1}{2}) = s_{\uparrow\uparrow k} \tag{9.3}$$
etc.

Having defined the spinor states for a particle in eqs. 9.1 and 9.2 we now introduce space and time, which of necessity must be in the knowledge base of Reality. Quite likely the three real space dimensions and the one real time dimension of our experience are the residue of a larger space-time. But they are what we experience so they must appear in the knowledge base—perhaps as a consequence of symmetry breaking.[120] So we introduce the coordinates $x = (t, \mathbf{x})$ where $\mathbf{x}$ is a 3-

---

[119] Again we choose not to deviate from our line of discussion to discuss the details of particle spin and refer the interested reader to popular books on elementary particles. The point we are seeking to establish is that the direct sum of the matrix representation of logical universes of discourse mathematically leads to the mathematics of spin ½ Dirac particles in a simple and direct manner. And that gives us the mathematical connection of Ideas to Reality of which Platonists spoke.

[120] Blaha (2008) proposes they are the result of the breaking of a 4 complex dimensional space-time. Superstring theories also propose mechanisms for the breakdown of theories of many dimensions to our known 4-dimensional space-time.

vector, and the momentum $p = (p^0, \mathbf{p})$ where p^0 is the energy and $\mathbf{p}$ the 3-momentum of a particle.

Then we form a Fourier representation of a spin ½ particle wave function:

$$\psi(x) = \sum_{\pm s} \int d^3p N(p)[b(p,s)u(p, s)e^{-ip\cdot x} + d^\dagger(p,s)v(p, s)e^{+ip\cdot x}] \qquad (9.4)$$

where $N(p)$ is a normalization factor, $b(p,s)$ is lowering operator that annihilates a particle of momentum p and spin s, $d^\dagger(p,s)$ is a creation operator, $u(p, s)$ and $v(p, s)$ are column vectors like those in eqs. 9.1 and 9.2 that have been "boosted" to momentum p, and $p\cdot x = p^0 x^0 - \mathbf{p}\cdot\mathbf{x}$ is the inner product of 4-vectors.[121] The raising and lowering operators $b(p,s)$ and $d^\dagger(p,s)$ (and their hermitean conjugates $b^\dagger(p,s)$ and $d(p,s)$) are mathematically similar to the raising and lowering operators of eq. 6.17. They satisfy the anticommutation relations

$$\{b(q,s), b^\dagger(p,s')\} = \delta_{ss'}\delta^3(\mathbf{q} - \mathbf{p}) \qquad (9.5)$$
$$\{d(q,s), d^\dagger(p,s')\} = \delta_{ss'}\delta^3(\mathbf{q} - \mathbf{p})$$

using Dirac δ-functions instead of Kronecker δ-functions due to the continuous nature of the momentum variables.

Thus we see spin ½ particle wave functions as emanating from the spinors, and raising and lowering operators, of the spinor formulation of Operator Logic.

Another interesting point that emerges from this discussion is the nature of spin ½ particle states such as

$$|p, s> = b^\dagger(p, s)|0> \qquad (9.6)$$

This state has an interpretation in reality as a one particle state. It also has an interpretation in Operator Logic as creating a one term universe of discourse which is in part linguistic and in part logic.

---

[121] This equation is discussed in detail in the many good books on quantum field theory. Again we wish to hold to our course of establishing the connection to Reality without deviating into digressions on the specifics of quantum field theory.

9.5.4 Particles as Monads

The combination of the physical interpretation and the Operator Logic interpretation is somewhat similar to Leibniz's concept of a *monad*. However it differs in that no spiritual aspect is evident[122] and also in that states, such as that defined in eq. 9.6, do not have "perception" in the usual sense of the word. But they do have something like perception through their interactions in the sense that they are affected by interactions, and thus may be said to 'perceive" other monads (particles) interacting with them.

9.5.5 Dirac-like Equations for Spin ½ Matter

Subsection 9.5.3 indicated in a general way the manner in which we can develop a Dirac-like wave function within the framework of the spinor matrix formulation of Operator Logic. In this section we will show how Dirac-like equations can be obtained by "boosting" the Dirac equation for a particle at rest to a particle of momentum $p = (p^0, \mathbf{p})$. The novel feature of this derivation is that the particle mass turns out to be a scale factor (with the dimension of [mass]) times a Gödel number.

Our starting point is A_{E_k} in eq. 6.16 which expanded to show its 4 × 4 rows and columns is

$$A_E = \begin{bmatrix} gn(E_1) & 0 & 0 & 0 \\ 0 & 0 & 0 & 0 \\ 0 & 0 & gn(E_2) & 0 \\ 0 & 0 & 0 & 0 \end{bmatrix} \tag{9.7}$$

where we surpress the subscript k. A_E can be put into a more physically relevant form using the unitary transformation[123]

$$U = \begin{bmatrix} 1 & 0 & 0 & 0 \\ 0 & 0 & 1 & 0 \\ 0 & 1 & 0 & 0 \\ 0 & 0 & 0 & 1 \end{bmatrix} = U^\dagger = U^{-1} \tag{9.8}$$

---

[122] A religious reader might say that since everything in Reality emanates from God an implicit spiritual aspect is present.

[123] This transformation poses no physics or logic issues since 4×4 Dirac γ matrices are equivalent up to a unitary transformation. See p. 18 of Bjorken (1965) and also R. H. Good Jr., Re. Mod. Phys. **27**, 187 (1955).

$$A_E' = UA_EU^{-1} = \begin{bmatrix} gn(E_1) & 0 & 0 & 0 \\ 0 & gn(E_2) & 0 & 0 \\ 0 & 0 & 0 & 0 \\ 0 & 0 & 0 & 0 \end{bmatrix} \tag{9.9}$$

If we now let $E_2 = E_1$ and $A_E' = A$ then

$$A = \begin{bmatrix} gn(E_1) & 0 & 0 & 0 \\ 0 & gn(E_1) & 0 & 0 \\ 0 & 0 & 0 & 0 \\ 0 & 0 & 0 & 0 \end{bmatrix} \tag{9.10}$$

$$= gn(E_1)(\gamma^0 + I_4) \tag{9.11}$$

where I_4 is the 4×4 identity matrix and γ^0 is one of the four Dirac γ matrices defined in eq. 6.20. If we multiply eq. 9.11 by a mass scale m_0 and define the particle mass[124]

$$m = gn(E_1)m_0 \tag{9.12}$$

then we have

$$m_0 A = m\gamma^0 + mI_4 \tag{9.13}$$

If we apply eq. 9.13 to a Dirac spinor U such as those in eqs. 9.1 or 9.2 (and an exponential factor) and set it equal to zero we obtain the Dirac equation for a particle at rest:

$$(m\gamma^0 + mI_4)Ue^{-imt} = 0 \tag{9.14}$$

---

[124] The reader may wonder whether the known quark and lepton masses – particularly their ratios have the Gödel number form of $2^a3^b5^c7^d...$ where a, b, c, d, ... are integers. Unfortunately the masses of most particles are not well known, and we expect the Higgs Mechanism and other mechanisms that may modify particle masses to ruin the simple Gödel ratios. The known mass ratios of the charged leptons are not exact Gödel ratios. There were observations of factors of 3 in the mass spectrum of "constituent" quarks some years ago with the mass ratios of s::c::b quarks being 3. $m_s = 0.5$ Gev, $m_c = 1.5$ Gev, and $m_b = 4.5$ Gev. (See "On A Possible Similarity Between The Heavy Lepton And Heavy Constituent Quark Mass Spectra", S. Blaha, Phys.Lett. **B84**, 116 (1979)) Since then the masses of these quarks have been adjusted to experimentally more acceptable values and the almost exact ratio of 3 has disappeared.

where t is the time variable. If we now perform a "Lorentz boost" – a Lorentz transformation to a reference frame moving with a velocity v with respect to the reference frame in which the particle is at rest, and described by eq. 9.14, we obtain[125]

$$(\not p - m)e^{-ip\cdot x}U(p) = 0 \qquad (9.15)$$

where the exponential factor, mt, is also boosted to p·x. The 4-momentum $p = (p^0, \mathbf{p})$ satisfies the mass condition

$$(p^0)^2 - |\mathbf{p}|^2 = m^2 \qquad (9.16)$$

where m is given by eq. 9.12. A Fourier transformation of eq. 9.15 yields the free, coordinate space Dirac equation:[126]

$$(i\gamma^\mu \partial/\partial x^\mu - m)\psi(x) = 0 \qquad (9.17)$$

with the general solution given by eq. 9.4. The Dirac equation, generalized to include internal symmetries associated with the Weak Interaction and Strong Interaction, incorporating the three known generations of quarks and leptons, and with interaction terms introduced, is the matter sector of the Standard Model of Elementary Particles.

Thus we have found a mathematical path from Operator Logic implemented by a spin ½ matrix formalism to the basic equations of matter. Plato, of course, had no knowledge of Operator Logic or The Standard Model or the mathematical path that we have outlined between them. But he and his successors were able to develop a conceptual outline of the realm of Ideas and a conceptual mathematical bridge to Reality.

Blaha (2008) developed a set of axioms that leads to the Standard Model. The steps from the Dirac equation above to the matter sector of the Standard Model may be briefly summarized:

---

[125] Blaha (2008) p. 29.

[126] Three other Dirac-like equations can be obtained using complex Lorentz boosts. The set of four equations corresponds naturally to the four known types of spin ½ particles and leads naturally to the form of the Standard Model.

1) The γ matrices being 4×4 matrices imply a 4-dimensional space-time (Weinberg (1995) p. 216).

2) Assume the complex Lorentz subgroup L_c of GL(4) is the space-time group of particles we find four types of spin ½ Dirac-like equations follow through boosts from the rest state to a state with one real time variable. These correspond to the four known types of spin ½ particles: up-quark, down-quark, electron-like and neutrino-like.

3) The four free Dirac-like equations are naturally grouped in doublets to obtain L_c covariant dynamic equations, and parity violation naturally appears as does color SU(3).

4) Upon the introduction of weak interaction coupling terms and particle mass splitting L_c is broken to Lorentz covariance. With the addition of another GL(4) group the three known generations of spin ½ particles and mass mixing follow yielding the known particle matter sector of the Standard Model. See Blaha (2008) for a detailed discussion.

10. Beyond the Standard Model

10.1 Approaches to a Deeper Level of Reality

While we have shown that Operator Logic furnishes a basis for the development of the Standard Model of Particles, the additional assumptions – what we call the Knowledge Base – remain to be determined. It may be that, like Euclidean geometry, we can go no further then a set of axioms (assumptions). The number of explicit axioms needed to obtain the Standard Model in Blaha (2008) was twenty-three. There were also a number of implicit assumptions in the derivation in Blaha (2008) just as there are a number of implicit assumptions in Euclid's'geometry which are only evident when figures are drawn in the process of proving theorems.

Irrespective of the question of the number and content of the axioms required to prove the Standard Model of Elementary Particles and thus establish the basis of Reality as we know it, the question that immediately arises is the source/reason for these particular axioms.

The possible sources for the axioms leading to the Standard Model are:

1. There is no reason. They are just the basis of Reality.
2. They follow from an unknown deeper unifying principle(s).
3. They follow from a deeper known theory such as Superstring Theory.
4. They follow from an unknown mechanism that establishes order in the form of the axioms from chaos. A mechanism or explanation for the persistence of such order over billions of years must also be found.

5. They follow because they are the only totally consistent set of physical axioms.
6. They are the result of chance and one of many possible sets of axioms.
7. They follow because they are required for life, as we know it, to exist. (The Anthropic Principle) This case is clearly a subcase of cases 1 and 6.

While case 1 is possible it is intellectually unsatisfying and so we will not pursue it. We will discuss case 2 in the following subsections. We will not discuss case 3 and refer the reader to the physics literature.

Case 4 seems unlikely to the author because of the secondary requirement that order persists indefinitely.

Case 5 brings us to Gödel's Consistency Theorem, which roughly states that a set of axioms cannot be proved to be consistent within the mathematical-deductive system that they define. Thus case 5 is unprovable.

Case 6 introduces an infinity of possible choices for the set of axioms for physics. The broadness of the set of choices makes this an unattractive possibility.

Case 7 is certainly true for humanity but recent studies have shown there is an extremely wide range of environments that can support life, and possibly, intelligent life. Thus, unless one posits religious reasons, it is not particularly clear that there is a compelling case for the Anthropic Principle. Case 7 then degenerates to case 6 – mere chance. And mere chance is not susceptible to intelligent discussion unless one can show that there is a "probability distribution" for sets of axioms and the set of axioms that governs our universe is amongst the most likely sets of axioms. Then case 7 becomes a subcase of case 2.

Therefore we find that case 2 is the most attractive case to study although case 3 (discussed elsewhere) is also of interest.

10.2 Extremum Rule for the Lagrangian of Physical Reality?

In Blaha (2005b) we developed a classification scheme for the "space" of all possible lagrangians and divided the space into subsets based on a definition of a Gödel number for lagrangians:[127]

> Definition: Suppose the Lagrangian L is an expression (string) of symbols, β_1, β_2, β_3, β_4, ... β_n (including field operator symbols, mathematical symbols, integral signs, the number of space dimensions, the number of time dimensions, space-time indices, and any other quantities that appear directly or indirectly in the Lagrangian); and v_1, v_2, v_3, v_4, ... v_n is the sequence of tokens corresponding to these symbols, then the Gödel number of a Lagrangian L is the integer
>
> $$gn(L) = \text{Min} \prod_{m=1}^{n} Pr(m)^{v_m}$$
>
> where $Pr(m)^{v_m}$ is the mth prime number raised to the power v_m with the 1st prime number taken to be 2; and where Min indicates the minimum is taken over all possible permutations of the order of symbols in the lagrangian that maintain its well-formed nature and all possible permutations of the assignment of tokens to the symbols in the lagrangian. The token numbers of the n symbols consist of a set of odd numbers ranging from $3 + p$ to $3 + p + 2(n - 1)$ where p is an even number greater than or equal to zero. (We set $p = 0$ for the sake of convenience.) *The set of token numbers consists of all odd integers ≥ 3.* A *well-formed lagrangian* has its symbols ordered in such a way that their order conforms to the syntax rules of the operators in the lagrangian. If L is an empty string (containing no symbols) then $gn(L) = 1$.

Since any number has a unique decomposition as a product of powers of prime numbers the following corollaries hold.

21. Corollary: If L_1 and L_2 are expressions such that $gn(L_1) = gn(L_2)$, then $L_1 = L_2$ (the Lagrangians are the same Lagrangian.)

22. Corollary: If the Gödel numbers of two Lagrangians are the same then they generate the same physical theory.

---

[127] Blaha (2005b) pp. 151-154.

If we take an arbitrary lagrangian and generate all possible lagrangians from it through substitutions, definitions of new variables (and field operators if the lagrangian is for a field theory), and other Rules of Inference that are used to transform lagrangians to a new form, then the set of lagrangians so constructed constitutes a set of physically equivalent theories. (Certain additional changes are made in the case of theories with path integral formulations.) As stated, we denote these sets of physically equivalent theories as LPset$_i$ for i = 1, 2, ... These sets are subsets of the set of all lagrangians Lset. The use of physically equivalent lagrangians is exemplified by a Higgs particle lagrangian, which is transformed to a different form to show the effects of its constant ground state value.

Those subsets of Lset, LPset$_i$, that contain classical (deterministic) physical theories such as theories in classical mechanics require no special treatment. Goldstein (1965) describes various forms of transformations: Legendre transformations and other transformations between equivalent physical theories.

In the case of subsets of Lset, LPset$_i$, that contain quantum theories, changes of variables (fields) require an additional modification in the path integral formulation of the theories. The path integral integrand must have a Jacobian factor that reflects the change of variables in the lagrangian in order for the original lagrangian theory and the transformed lagrangian theory to have the same physical implications. For example if the path integral for a certain lagrangian in a certain subset of lagrangian theories is

$$Z(J) = N \int D\phi \, \exp\{i \int d^4 y \, [\mathscr{L}(\phi) + J(y)\phi(y)]\}$$

where N is a normalization factor. Then a change of field variable from ϕ to ψ

$$\phi = \phi(\psi)$$

in the lagrangian necessitates the introduction of a Jacobian factor $\mathscr{J}(\psi)$ in the path integral integrand (except for linear changes of variable)

$$Z(J) = N \int D\psi \, \mathscr{J}(\psi) \exp\{i \int d^4 y \, [\mathscr{L}(\psi) + J(y)\phi(\psi(y))]\}$$

See Huang (1965) or Weinberg (1998) for examples in the case of Fadeev-Popov gauge fixing.

Thus we have shown that it is possible to define subsets of theories for deterministic and quantum lagrangians where each lagrangian in the subset embodies the same physical theory.

We can define a Gödel number for a subset LPset$_k$ as:

23. Definition: The Gödel number of a subset LPset$_k$ is the minimum of the Gödel numbers of the members of the subset.

Due to the unique decomposition of a whole number into its prime factors we have the lemma:

24. Lemma: If the Gödel numbers of two subsets of Lpset are the same then they are the same subset.

In the next part we will define a criteria for selecting the "simplest" lagrangian in a given subset. This definition will be of some importance since potential Theories of Everything are judged by their simplicity as well as their physical implications.

The criteria of simplicity and elegance, which are much talked about in discussions of the Theory of Everything, are subjective, ambiguous, and anthropomorphic. The assignment of Gödel numbers to lagrangians enables us to create a mathematical definition of simplicity and, to some extent, of elegance since these criteria are usually subjectively related. A lagrangian's Gödel number is a measure of the number of symbols and the number of times each symbol appears in the lagrangian. Therefore lagrangians in fewer dimensions, with fewer species of particles or fields, with lower spin entities, and with fewer terms will have a smaller Gödel number. Based on this observation we define the simplicity of a lagrangian as:

25. Definition: For any two lagrangians L_1 and L_2 if $gn(L_1) < gn(L_2)$ then L_1 is simpler than L_2.

which implies the following lemma:

26. Lemma: The Gödel number of the subset LPset$_{TE}$ that contains the Theory of Everything is the Gödel number of the simplest lagrangian in the subset.

The preceding quote gives us a method of finding the Gödel number of the set of equivalent lagrangians that define the "Theory of Everything". However an extremum method for determining this set of lagrangians is lacking. Clearly it is not defined by the least Gödel number. And, in fact, there does not appear to be an obvious connection between Gödel numbers and lagrangians that would lead to an extremum method.

Leibniz, who invented calculus including extending extremal (minimax) methods,[128] suggested perhaps the most concrete concept for an extremum rule to determine the fundamental laws of nature: *the laws of nature are of maximal simplicity yet capable of producing a universe of maximal complexity.*[129]

The union of these two features was one of the motivations that led Leibniz to describe this universe as "the best of all possible worlds." The mathematical framework needed to express this elegant, but currently unquantifiable, extremum principle, or any extremum principle that leads to the laws of space, time and matter, remains to be developed. One can hope that someday we will reach that level of understanding if finding the "right" lagrangian with this approach is the correct approach. It seems difficult to see how this approach can determine the features of space-time and of internal symmetries and interactions.

If we do, then we will perhaps be able to create "universes of the mind" in which we can explore the implications of other designs for a universe. And perhaps we may be able in that latter age to experimentally create bubbles of those other possibilities for a universe in the laboratory.

Whether there may be other sister universes that realize these other designs is also a question of importance. Today physicists speculate about such universes but if the history of physics has taught us anything, it has taught us that very few speculations about future physical theory turn out to be correct.

10.3 Axiomatic Approach to the Knowledge Base of Reality

Blaha (2008), and his earlier books, developed an axiomatic approach that led *exactly* to the established parts of the lagrangian of the Standard Model of Particles. In this approach an axiomatic method was used and the established parts of the Standard Model derived.

---

[128] Nova Methodus pro Maximis et Minimis ("New Method for the Greatest and the Least") 1684.
[129] In Leibniz's words, "at the same time the simplest in hypotheses [i.e., its laws] and the richest in phenomena." Quoted in Rescher (1967) p.19.

The success of this approach leads us to consider physics to ultimately be the result of a set of fundamental axioms that lead to the Standard Model lagrangian. It is possible to define Gödel numbers for a set of axioms in a manner similar to that of the preceding subsection for lagrangians. Blaha (2005b) defined Gödel numbers for sets of axioms:[130]

27. Definition: If A is a non-Lagrangian, axiomatic theory, and a_1, a_2, a_3, ... , a_n are the *sequences* of symbols of the n (different) axioms of A, then the Gödel number of the theory can be defined to be

$$gn(A) = Min \prod_{m=1}^{n} Pr(m)^{gn(a_m)}$$

where $Pr(m)^{gn(a_m)}$ is the mth prime number raised to the power $gn(a_m)$, where $gn(a_m)$ is the Gödel number of the axiom a_m and where Min indicates the minimum is taken over all possible permutations of the order of the axioms, all possible orderings of the symbols in the axioms that result in well-formed axioms, and all possible permutations of the assignment of tokens to the symbols in the production rules.

Thus if there are n axioms rules that use m symbols, then there at least n!m! Gödel numbers that correspond to the production rules (assuming the token numbers consist of the set of odd numbers ranging from $3 + k$ to $3 + k + 2(m - 1)$ where k is an even number greater than or equal to zero. (We set $k = 0$ for the sake of convenience.) A unique Gödel number for A is obtained by taking the prescribed minimum.

28. Theorem: If the Gödel numbers of two axiomatic theories are the same then they generate the same physical theory.

15.6.2 Subsets of NLset Containing Physically Equivalent Theories

Given a set of primitives and a set of axioms that define a physics theory, we can create a mathematical-deductive system that describes the physical theory in a manner similar to that used to define a mathematical-deductive system for a mathematics theory such as the theory of whole numbers.

Suppose we now take the primitive terms p_1, p_2, ... of the theory A and define a new set of primitive terms p_i' with

---

[130] Blaha (2005b) pp. 154-156.

$$p_k = p_k(p')$$

and substitute the new primitives in the axioms creating a new theory A'. The new theory is physically equivalent to the original theory A but its mathematical-deductive system will look different.

Now consider the set of physical theories generated by all possible redefinitions of the primitive terms as above. This set of physical theories will be a subset of the set of all non-Lagrangian theories NLset. All theories in this subset are physically equivalent.

A moment's consideration leads to the realization that NLset is composed of subsets of equivalent axiomatic theories which we will denote $NLset_i$ for $i = 1, 2, \ldots$ The elements of each subset have different Gödel numbers in general. Therefore we will define the Gödel number of a subset by:

29. Definition: The Gödel number of a subset $NLset_k$ is the minimum of the Gödel numbers of the members of the subset.

Due to the unique decomposition of a whole number into its prime factors we have the lemma:

30. Lemma: If the Gödel numbers of two NLset subsets are the same then they are the same subset.

In the next section we will define a criteria for selecting the "simplest" axiomatic theory in a given subset.

15.6.3 Simplicity Criteria for Subsets of NLset Containing Physically Equivalent Theories

A Gödel number is a measure of the number of symbols and the number of times each symbol appears in the set of axioms. Therefore sets of axioms for physics theories in fewer dimensions, with fewer species of particles or fields, with lower spin entities, and with fewer terms will have a smaller Gödel number. Based on this observation we define the simplicity of an axiomatic theory by:

31. Definition: For any two axiomatic theories A_1 and A_2 if $gn(A_1) < gn(A_2)$ then A_1 is simpler than A_2.

which implies the following lemma:

32. Lemma: The Gödel number of the subset $NLset_{TE}$ that contains the Theory of Everything, if it is axiomatic in nature, is the Gödel number of the simplest theory in the subset.

15.7. Escaping the Trap of Anthropomorphism

The form of Theories of Everything, and of physical theories in general, normally has anthropomorphic trappings: the choice of symbols, the form of the expressions, the characterization of the symmetries, and so on.

If we use the Gödel-number-based approach to defining lagrangian and non-lagrangian theories using the definitions of simplicity that we have developed we arrive at a non-anthropomorphic characterization of physical theories.

Thus we can hope that we can arrive at a universal characterization of the Theory of Everything that would pass this test: if we meet an intelligent alien scientist we could compare our theories of everything based on Gödel numbers! If the theories are the same their Gödel numbers should be equal. (We assume that they define tokens in the same way.)

As the passage shows one can define Gödel numbers for sets of axioms as well as for lagrangians. Unfortunately the same issue arises: there is no principle(s) that can be used to select the set of axioms actually used by Nature. Blaha's axioms are quite simple, relatively speaking, and lead directly to the Standard Model of Particles. It seems the best approach to progress in this area is to refine, and to reduce the number of, this set of axioms as well as to develop physical motivations for the axioms. If progress is made in this area then perhaps the Gödel number classification scheme may prove of interest.

11. Logic and Language

11.1 Logical Equivalence of Languages

All logic is expressed in human or symbolic languages. As a result there is an intimate connection between logic and language. Statements and deductive systems require a sufficiently robust language to express their content. A language usually must have the equivalent of predicates, subjects, connectives and quantifiers.

A question of some interest is the equivalence of languages. When are two languages equally capable of expressing the statements of a universe of discourse? Clearly they must have sets of equivalent terms although a term in one language might be a combination of terms in the other. However in some cases of human languages this type of equivalence is hard to achieve. A classic example is the Greek language in comparison to English. The Greek language has hundreds of words expressing various forms and nuances of love while English, in comparison, has few words for love. In other areas of the world some languages have a plethora of words for one aspect of nature, or another, which have no simple analogue in European languages.

One might think that the equivalence of languages is not of great importance. However, the growth of culture and science is directly tied to the growth in their terminology and the concepts that they embody. An example is the growth in the knowledge of quantum theory in the twentieth century, which introduced a host of new terms in physics. Thus the equivalence of languages reflects, to some degree, the equivalence of the range of universes of discourse (and their intellectual content) that the languages can support.

The requirements for two languages to be equivalent are:

1. Equivalent expressions in the two languages must have the same truth value.

2. Any expression in one language must have an expression with equivalent meaning in the other language.

3. The primitive terms of one language must be equivalent to the primitive terms in the other language, or to combinations of the primitive terms in the other language.

11.2 Operator Languages

Hitherto the languages that have been considered in studies of Logic have been human languages or symbolic languages that we will call c-number languages following the terminology of Quantum Theory. A c-number is a quantity, a number, character, or string of characters, that is not an operator. We have introduced the concept of a q-number language for logic in the form of Operator Logic and Quantum Operator Logic. Primitive terms are represented by operators in a Hilbert space. This q-number language is as valid as c-number languages to express statements.

Q-number languages have the added advantages of

1. Being able to project out undecidable statements and thus resolving the profound issues raised by the Gödel Undecidability Theorem. C-number formulations of logic are an inadequate framework. Q-number logic is the proper framework for Logic.

2. Enabling the creation of subuniverses of discourses, direct products of universes of discourse, and direct sums of universes of discourse.

3. Furnishing a unifying framework for deterministic logic and quantum probabilistic logic that is guaranteed to be well-formed and consistent since it is based on Quantum Measurement Theory, which, embodying Reality, cannot be inconsistent or incomplete.

4. Providing matrix formulations of Operator Logic and Quantum Operator Logic.

5. Implementing the Platonic concept of Ideas connected mathematically to Reality.

11.3 Quantum Languages, Grammar, Turing Machines, Computers, and Computer Programs

This book develops Operator Logic and Quantum Operator Logic. Blaha (2005b) developed the concepts of Quantum Languages, Quantum Grammars, Quantum Turing Machines, Quantum Computers, and Quantum Computer Programs; and proved Gödel's Undecidability Theorem required the fundamental laws of Nature to be quantum.

These books complement each other by bringing the Quantum concept, which is undoubtedly the deepest knowledge that we have of Reality, to logic and language.

A reading of Blaha (2005b) displays a remarkably similarity in the concepts and mathematics of quantum languages with our present development of Quantum Operator Logic.

Together they give us a coherent weltanschauung of Thought and Reality. The major open question is the determination of the knowledge base of Reality. In this author's view this question will be resolved by an extension of our understanding of the formulation of the nature of space, time, and substance within the framework of Quantum Operator Logic and Quantum Language.

Appendix A. The Observer in Quantum Theory & Operator Logic

In Quantum Measurement Theory there is an implicit Observer who "sets up" experiments, performs eigenvalue measurements (filters) at various stages in an experiment, and measures the end result(s) of an experiment.

A.1 The Observer in Quantum Measurement Theory

The role of the Observer in Quantum Measurement Theory is extremely significant. One indication of the importance of the Observer in quantum theory is the effect of an observation (without filtration) of the state of an intermediate stage of an experiment compared to non-observation of the state of the intermediate stage of an experiment.

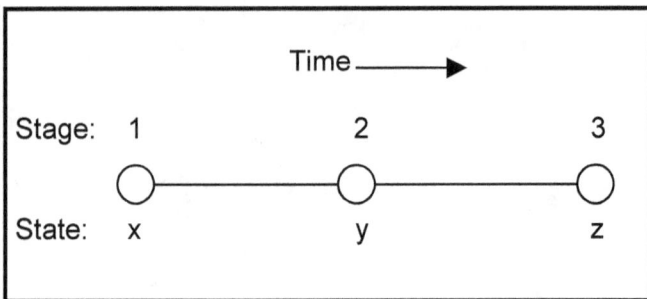

Figure A.1. A three stage experiment beginning with a system in the state with eigenvalue x of observable X, having an intermediate state 2 with the eigenvalue y of observable Y, and ending in state 3 with eigenvalue z of observable Z. Time proceeds from left to right in the diagram.

Consider the experiment diagrammed in Fig. A.1. The probability amplitude of the transition from eigenvalue x of X to eigenvalue y of Y and thence to eigenvalue z of Z is

$$<Z(z)|Y(y)><Y(y)|X(x)>$$

and thus the corresponding probability is

$$|<Z(z)|Y(y)>|^2|<Y(y)|X(x)>|^2 \tag{A.1}$$

If we do not filter at stage 2 of the experiment then we must sum over the Y eigenvalue states to obtain the probability of the transition from x to z:

$$p_Y(z, x) = \sum_y |<Z(z)|Y(y)>|^2|<Y(y)|X(x)>|^2 \tag{A.2}$$

$$= \sum_y <Z(z)|Y(y)><Y(y)|Z(z)><Y(y)|X(x)><X(x)|Y(y)> \tag{A.3}$$

Note the probability $p_Y(z, x)$ is dependent on the summation over intermediate (stage 2) Y eigenstates in the particular way specified by eq. A.3.

If stage 2 were not present in the experiment and the experiment ran from stage 1 directly to stage 3 without intermediate observation then a different probability, which we will denote p(z, x), would result.

$$p(z, x) = |<Z(z)|X(x)>|^2 \tag{A.4}$$

$$= \sum_{y_1} <Z(z)|Y(y_1)><Y(y_1)|X(x)> \sum_{y_2} <X(x)|Y(y_2)>< Y(y_2)|Z(z)>$$

$$= \sum_{y_1, y_2} <Z(z)|Y(y_1)><Y(y_1)|X(x)><X(x)|Y(y_2)>< Y(y_2)|Z(z)> \tag{A.5}$$

which is manifestly different from $p_Y(z, x)$ as given in eq. A.3.

Thus we conclude that the mere presence of an observation point (stage 2 in the present example) changes the probability for the experiment even if a filtration is not made.

We thus have concrete proof (substantiated by extensive experiments) that the passive observation of the state of an experiment in progress affects the probability of the outcome of the experiment.

Since an observation (passive, or active (i.e. a filtration)) necessarily requires an observer we see that observers influence the evolution of experiments, and thus have an important role in quantum theory. Since we base Operator Logic on Quantum Measurement Theory we conclude that a parallel observer role exists implicitly for Operator Logic.

A.2 The Observer in Operator Logic

In section 1.3 we considered many of the aspects of a statement. In particular we considered the sense of a statement. Implicitly the sense of a statement (and Frege's bedeutung) of a statement and its parts introduce the Observer into Logic and, more importantly, into Operator Logic.

At the simplest level the observer determines the truth (semantic universe of discourse) or provability (calculus universe of discourse) of a statement.

Actually the observer has three major roles in Operator Logic:

1. Create statements.
2. Evaluate the truth (provability) of statements.
3. Create derivations of statements (theorems).

The effect of the observer on the truth (provability) of statements depends on whether the statement is deterministic (non-quantum) or a quantum Operator Logic statement. In the case of deterministic Operator Logic the observer has no effect on the outcome (true or false, provable or not) of the statement. In the case of Quantum Operator Logic, if the observer senses the state of a statement at an intermediate point(s) of the statement then the probability of the statement is affected in the same manner as we saw in the previous section for Quantum Measurement Theory.

Appendix B. Time and Space for Reality and Operator Logic

In section 9.5 and particularly in subsection 9.5.5 we developed the beginnings of the derivation of the Standard Model of Elementary Particles. In doing so we introduced time and space and proceeded to obtain the Dirac equation for fundamental spin ½ particles (the basis of all known matter – Dark Matter is a separate unresolved issue experimentally and theoretically). This development was extended in Blaha (2008) to a complete derivation of the Standard Model.

The extension of Operator Logic to include time and space requires justification. How and why are time and space united with Operator Logic?

B.1 The Necessity of Time, and an Arrow of Time, in Operator Logic

A number of logicians have noted that a concept of time is implicit in conventional Logic. For example, proofs of theorems proceed step by step from initial postulates and theorems to a theorem's proof. Embedded in that process is a notion of discrete time steps, and a direction of the time steps. The directionality of the time steps[131] specifies an "arrow of time." The question of the Arrow of Time – why time proceeds forward and not backward – has been a subject of much discussion over the years. In the present situation Logic, and Operator Logic, automatically embody an arrow of time.

Not only is this true for proofs but it is also true for statements. Although the order of the parts of a statement are language dependent the order is specific and consecutive within a given language and thus has a time order as well.

---

[131] After all one does not proceed "backward" from a theorem through the proof steps to the initial postulats.

So we conclude that Operator Logic embodies discrete time steps and a definite concept of time ordering – an Arrow of Time.

Having ascertained that discrete time, and time ordering, is implicit in Operator Logic we now define physical time as the continuous limit of discrete time with the understanding that physical time may be discrete and may consist of very small time steps of time intervals of the order of the Planck time scale 5.39×10^{-44} seconds. Discrete time intervals of that order of magnitude are not detectable experimentally at present or in the foreseeable future. Thus the assumption of continuous time, with an arrow of time, is satisfactory.

B.2 Why add Spatial Dimensions to Operator Logic?

Spatial dimensions must be added to establish a connection with the spin ½ matrix formulation of Operator Logic, which we connected to the Dirac equation in section 9.5.5. This leads us to consider the connection between Operator Logic spinors and spinors in physical Reality. We therefore assume that there is a map between Operator Logic spinors and the physical spinors of fermions.

Clearly if spinors exist in the real world, as they do, then they must be "spinning" in spatial dimensions. The number of components of a spinor is related to the total number of time and space dimensions.[132] For the case of an even number of dimensions d a spinor has $2^{d/2}$ components. For the case of an odd number of dimensions d a spinor has $2^{(d-1)/2}$ components. Based on these formulas we find the results in the following table.

Total Number of Space-Time Dimensions d	Number of Spinor Components
1	1
2	2
3	2
4	4

Table B.1. The number of spinor components for various space-time dimensions.

---

[132] Weinberg (1995) p. 216.

The case of d = 1 is immediately ruled out because Operator Logic supports at minimum 2 component spinors or 4 component spinors. The case d = 2 is also ruled out because spinor particles in a one-dimensional space reduce to scalar particles, and Reality has true spinors. The case d = 3 is ruled out because in two spatial dimensions there is no difference between left-handedness and right-handedness. Thus the minimal number of spatial dimensions that yield true physical spinors and support "handedness" is three spatial dimensions. This case meets Leibniz's criteria of laws of maximal simplicity that produces a universe of maximal complexity. The simplest features associated with space are spin (represented by spinors) and handedness. They yield a rich spectrum of particle types and interaction types (maximal complexity).

Thus we have a rationale for the extension of Operator Logic to include one time and three spatial dimensions.

Appendix C. Being (Existence)

The question of being or existence has been a subject of discussion in Philosophy and Metaphysics for millenia. In the absence of "experimental" information the discussions have centered on the definition of being and the implications of these definitions for the "properties" of being. For many scholars the state of Philosophy and Metaphysics was considered satisfactory in the 20th century. For example, Hans Bethe, perhaps the dominant figure in theoretical physics from the 1930's through the 1950's, and a Nobel Prize winner, stated that at the beginning of his "graduate" studies in the mid-1920's he considered the state of Philosophy and Metaphysics, and concluded that they were satisfactory. He then decided to become a physicist where he felt that he could make significant contributions (which he did by discovering the solar energy carbon cycle, and making notable contributions to quantum field theory as well as guiding a generation of physicists including R. P. Feynman and M. Gell-Mann).

In this appendix we will consider a phenomenological view of being and nothingness based on the experimental observation of the creation and annihilation of particles. The experimental picture that we will use dates from the 1930's when the occurrence of particle creation and annihilation was first recognized. The mathematics of quantum field theory adequately describes particle creation and annihilation as seen experimentally. Since particle creation is the "creation" of being and since particle annihilation is the "destruction" of being one could say that the issues of being and non-being can be resolved by experimental observations and their theoretical analysis.

However one also recognizes that the mathematics of creation and annihilation somehow doesn't fully answer the question, What is being? Part of the problem is that we don't exactly know what a particle is. We know particles have particle-like properties, and wave-like properties as well. But can we say particles are composed of a substance? If so, what substance? Or are particles merely form without substance?

C.1 The Substance of Particles

Philosophers have long argued (two and a half millenia) about the nature of matter and energy. These discussions have largely centered around the definition of these entities followed by an analysis based on one definition or another. The question of their reality was also a recurring issue. In the earliest discussions the world was thought to consist of varying combinations of the four elements: earth, air, fire and water. So early philosophical discussions took the lead from "experimental" observation because those four elements are what we encounter in nature, and one can argue that various materials seen in nature are combinations of the four elements. More recently philosophers such as Bishop Berkeley have raised the question whether the world is real or perhaps some form of evolving "thought" in the mind of God, or a similar insubstantiality.

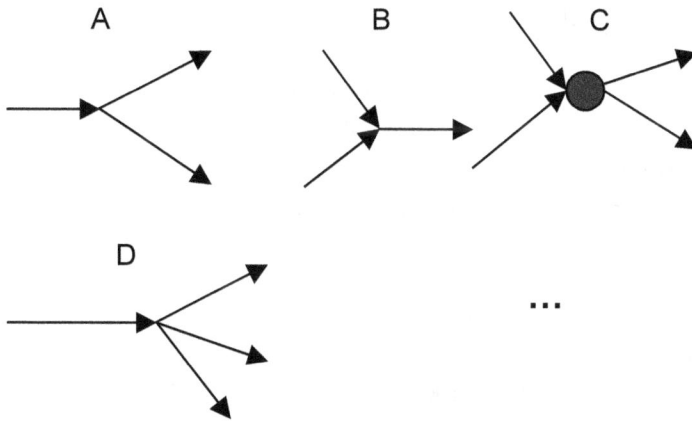

Figure C.1. Some diagrams showing how particles interact with each other and transform between types of particles. In diagram A a particle "splits" into two particles. In B two particles combine to produce an "outgoing" particle. In C two particles interact in some complex way represented by the filled circle to produce two outgoing particles which may or may not be the same particles. In D a particle transforms (decays) into three particles.

In this appendix we address the issue of being based on experimental observations – particularly particle observations made from the 1930's onward. *We look to experiment/reality as the first philosophers did to uncover the nature of matter and energy.*

The first experimental reality that we encounter is that elementary particles can transform into each other in certain ways that are theoretically well described by the Standard Model of Elementary Particles.[133] Some typical transformations are diagrammed in Fig. C.1.

Because of the nature of the Standard Model, particles of matter (quarks and leptons) and particles of energy (photons and massive vector bosons) can transform into each other in all sorts of ways if energy and momentum are conserved, and if the incoming particle(s) have sufficient energy to generate the outgoing particles.

Since, under the right energy and momentum conditions, it is possible for any incoming set of particles to transform into any outgoing set of particles[134] it seems reasonable to assume that the substance of all particles in nature is the same, and that particles differ only in their form. This assumption reflects the principle called Ockham's Razor[135] which states that usually the simplest explanation of a phenomena is the correct one.[136]

So instead of assuming each particle is composed of a different substance the well-substantiated Standard Model containing transformations (interactions) between all the known particles suggests that particles are made of the same substance (or perhaps no substance) and their differences are due to their internal form. There is no experimental data that could clarify the nature and properties of this substance. There is a further problem in that the dominant form of matter in our universe is Dark Matter and the dominant form of energy is Dark Energy. Dark Matter and Dark Energy interact with "normal" matter and

---

[133] There are some uncertainties with regard to a small number of special transformations and the possibility of new particles and new types of transformations at higher energies is not excluded. But the overwhelming majority of particle physics phenomena are well described by the Standard Model.

[134] With the proviso that charge and internal quantum numbers such as strangeness, color, and so on are properly conserved (or not conserved according to the dictates of the Standard Model).

[135] William of Ockham – Law of Parsimony – "Pluralitas non est ponenda sine necessitate" or "Plurality should not be posited without necessity." First stated by Durand De Saint-Pourçain (1270-1334 AD).

[136] Richard P. Feynman expressed a similar view – that the simplest solution is often the correct one.

energy through the gravitational interaction, which is very weak. So the substance of which Dark Matter and Dark Energy are made are open questions.

It is possible that all types of matter and energy are not made of any substance but instead consist of form imposed on nothingness. Then Dark Matter and Dark Energy would be "dark" only because they did not interact with normal matter and energy except through the very weak force of gravity.

It is difficult to imagine form imposed on nothingness yet this possibility is not ruled out. There is also the question of how nothingness relates to the vacuum. In quantum field theory the vacuum is a quantum thing and possesses properties. It possesses enormous energy. Vacuum fluctuations can occur in which particle – antiparticle pairs "pop out" of the vacuum for extremely short periods of time. The vacuum can "exert" Casimir forces that can be measured in the laboratory. So the vacuum is an extremely dynamic "substance" in quantum field theory.

If we identify the vacuum of quantum field theory with nothingness then nothingness is an extremely dynamic thing as well. Perhaps its major distinguishing feature then becomes its non-conservation. Nothingness is unlimited in quantity as opposed to substances of everyday experience which have a fixed mass or size. One thinks of the sacred law of 19th century physics, "Mass is conserved and cannot be created or destroyed," which was overturned by Einstein's theory of Relativity and the transformation of matter into energy in fusion and fission. Nothingness being unlimited can easily be interpreted as the substance of particles and its unlimited nature enables transitions between the numbers and types of particles.

C.2 The Form of Particles

If as the previous section suggested particles are composed of one substance, and particles are differentited from one another by their form, then the connection of the realm of Ideas with Reality becomes less difficult since form is based on Ideas – theoretical concepts. We have learned that the form of particles is in part based on space-time features such as their spin, their energy, their mass, and their momentum. In addition, their form is based on their electric charge and their internal quantum numbers. These quantities are determined by their group structure. Their group structure is $SU(2){\otimes}U(1)$ (ElectroWeak symmetry) and color SU(3) (quark and gluon particle Strong interaction symmetry). There may be additional symmetries that will appear in the future as higher energy accelerators discover new phenomena.

But, from what we have learned, it is clear that space-time symmetry and internal group symmetries determine the form of particles. (See Blaha (2008) for an axiomatic derivation of these symmetries and the form of the Standard Model.)

C.3 Being as Form

The consequences of the considerations in the previous two sections suggest that being is form and there is one unlimited substance in nature from which the particles of Reality and thus Reality is constructed.

If being is form, then there is no difficulty in adding being to Operator Logic to connect it to Reality using the Knowledge Base of space-time features and internal symmetries.

Thus we have realized the Platonic scheme of a realm of Ideas connected mathematically to Reality.

REFERENCES

Bjorken, J. D., Drell, S. D., 1965, *Relativistic Quantum Fields* (McGraw-Hill, New York, 1965).

Blaha, S., 2004, *Quantum Big Bang Cosmology: Complex Space-time General Relativity, Quantum Coordinates,*™ *Dodecahedral Universe, Inflation, and New Spin 0, ½, 1 & 2 Tachyons & Imagyons* (Pingree-Hill Publishing, Auburn, NH, 2004).

Blaha, S., 2005a, Quantum Theory of the Third Kind: A New Type of Divergence-free Quantum Field Theory Supporting a Unified Standard Model of Elementary Particles and Quantum Gravity based on a New Method in the Calculus of Variations (Pingree-Hill Publishing, Auburn, NH, 2005).

Blaha, S., 2005b, *The Metatheory of Physics Theories, and the Theory of Everything as a Quantum Computer Language* (Pingree-Hill Publishing, Auburn, NH, 2005).

Blaha, S., 2005c, *The Equivalence of Elementary Particle Theories and Computer Languages: Quantum Computers, Turing Machines, Standard Model, Superstring Theory, and a Proof that Gödel's Theorem Implies Nature Must Be Quantum* (Pingree-Hill Publishing, Auburn, NH, 2005).

Blaha, S., 2006, *A Derivation of ElectroWeak Theory based on an Extension of Special Relativity; Black Hole Tachyons; & Tachyons of Any Spin.* (Pingree-Hill Publishing, Auburn, NH, 2006).

Blaha, S., 2007b, *The Origin of the Standard Model: The Genesis of Four Quark and Lepton Species, Parity Violation, the ElectroWeak Sector, Color SU(3), Three Visible Generations of Fermions, and One Generation of Dark Matter with Dark Energy* (Pingree-Hill Publishing, Auburn, NH, 2007).

Blaha, S., 2008, *A Complete Derivation of the Form of the Standard Model With a New Method to Generate Particle Masses SECOND EDITION* (Pingree-Hill Publishing, Auburn, NH, 2008)

Braithwaite, R. B., 1960, *Scientific Explanation* (Harper Torchbook, New York, 1960).

Carnap, R., 1956, *Meaning and Necessity* (Univ. Chicago Press, Chicago, 1956).

Carnap, R., (Ed. M. Gardner), 1995, An *Introduction to the Philosophy of Science* (Dover Publications, New York, 1995).

Curry, H. B., 1976, *Foundations of Mathematical Logic* (Dover Publications, New York, 1976).

Davis, M., 1982, *Computability and Unsolvability* (Dover Publications, New York, 1982).

Dirac, P. A. M., 1931, *Quantum Mechanics* Third Edition (Oxford University Press, Oxford, 1947).

Frege, G., (Ed. M. Beaney), 1997, *The Frege Reader* (Blackwell Publishing, Malden, MA, 1997).

Garson, J. W., 2006, *Modal Logic for Philosophers* (Cambridge University Press, Cambridge, 2006).

Gödel, K., 1992, Tr. Meltzer, B., Introduction by R. B. Braithwaite, *On Formally Undecidable Propositions of Principia Mathematica and Related Systems* (Dover Publications, New York, 1992).

Gottfried, K., 1989, *Quantum Mechanics I Fundamentals* (Addison-Wesley, Reading, MA, 1989).

Hilbert, D. and Ackermann, W. (Tr. L. M. Hammond et al), 1950, *Principles of Mathematical Logic* (Chelsea Publishing Co., New York, 1950).
Kleene, S. C., 1967, *Mathematical Logic* (Dover Publications, New York, 1967).

Konyndyk, K., 1986, *Introductory Modal Logic* (University of Notre Dame Press, Notre Dame, Indiana, 1986).

Lavine, S., 1994, *Understanding the Infinite* (Harvard University Press, Cambridge, MA, 1994).

Mackey, G. W., 1963, Mathematical Foundations of Quantum Mechanics (W. A. Benjamin, New York, 1963).

Messiah, A., 1965, *Quantum Mechanics* Volume I (John Wiley & Sons, New York, 1965).

Potter, M., 2004, *Set Theory and its Philosophy* (Oxford University Press, Oxford, 2004).

Quine, W. van O., 1962, *Mathematical Logic* (Harper Torchbooks, New York, 1962).

Rescher, N., (1967), *The Philosophy of Leibniz* (Prentice-Hall, Englewood Cliffs, NJ, 1967).

Révész, G. E., 1983, *Introduction to Formal Languages* (Dover Publications, New York, 1983).

Smullyan, R. M., 1995, *First-Order Logic* (Dover Publications, New York, 1995).

Tarski, A., 1995, *Introduction to Logic and to the Methodology of Deductive Sciences* (Dover Publications, New York, 1995).

Tiles, M., 1989, *The Philosophy of Set Theory* (Dover Publications, New York, 1989).

Weinberg, S., 1995, *The Quantum Theory of Fields Volume I* (Cambridge University Press, New York, 1995).

Weyl, H., 1950, *Space, Time, Matter* (Dover, New York, 1950).

Weyl, H., (Tr. S. Pollard et al), 1987, *The Continuum* (Dover Publications, New York, 1987).

Index

About the Author

Stephen Blaha is an internationally known physicist with extensive interests in Science, the Arts, and Technology. He received his Ph.D. in Theoretical Physics from Rockefeller University (NY). He has written a highly regarded book on physics, consciousness and philosophy – *Cosmos and Consciousness*, a book on Science and Religion entitled *The Reluctant Prophets*, a book applying physics concepts to the history of civilizations, and books on Java and C++ programming. He developed a mathematical theory of civilizations that is described in *The Life Cycle of Civilizations*. Recently he completed a major new study of Cosmology: *Quantum Big Bang Cosmology: Complex Space-time General Relativity, Quantum Coordinates, Dodecahedral Universe, Inflation, and New Spin 0, ½, 1 & 2 Tachyons & Imagyons*. He has served on the faculties of several major universities. He was an Associate of the Harvard Physics Faculty for twenty years (1983-2003). He was also a Member of the Technical Staff at Bell Laboratories, a member of management at the Boston Globe Newspaper, a Director at Wang Laboratories, President of Blaha Software Inc and Janus Associates Inc. (NH), and 2008 Program Chair of the International Society for the Comparative Study of Civilizations.

Among other achievements he was a co-discoverer of the "r potential" for heavy quark binding developing the first (and still the only demonstrable) non-abelian gauge theory with an "r" potential; first suggested the existence of topological structures in superfluid He-3; first proposed Yang-Mills theories would appear in condensed matter phenomena with non-scalar order parameters; first developed a grammar-based formalism for quantum computers and applied it to elementary particle theories; first developed a new form of quantum field theory without divergences (thus solving a major 60 year old problem that enabled a unified theory of the Standard Model and Quantum Gravity without divergences to be developed); first developed a formulation of complex General Relativity based on analytic continuation from real space-time; first developed a generalized non-homogeneous Robertson-Walker metric that enabled a quantum theory of the Big Bang to be developed without singularities at t = 0; first generalized Cauchy's theorem and Gauss' theorem to complex curved multi-dimensional spaces; first developed a physically acceptable theory of faster-than-light particles – tachyons – of any spin; first showed a universe with three complex spatial dimensions has an icosahedral symmetry; first developed the form of the composition of extrema in the Calculus of Variations; first suggested that inflationary periods in the history of the universe were not needed; first proved Gödel's Theorem implies Nature must be quantum, first derived the form of the Standard Model, first developed a quantitative harmonic oscillator-like model of the life cycle, and interactions, of civilizations, and first developed an axiomatic derivation of the Standard Model.

Blaha was also a pioneer in the development of UNIX for financial and scientific applications, database benchmarking, and networking (1982); in the development of Desktop Publishing (1980's); and developed a hybrid shell programming technique (1982) that was a precursor to the PERL programming language. He received Honorable Mention in the Gravity Research Foundation Essay Competition in 1978, and was nominated for three "Awards for Technical Excellence" in 1987 by PC Magazine for PC software products that he designed and developed.